CAMBRIDGE COUNTY GEOGRAPHIES
General Editor: F. H. H. Guillemard, M.A., M.D.

OXFORDSHIRE

Cambridge County Geographies

OXFORDSHIRE

by

P. H. DITCHFIELD, M.A., F.S.A.

With Maps, Diagrams and Illustrations

Cambridge:
at the University Press
1912

CAMBRIDGE UNIVERSITY PRESS
Cambridge, New York, Melbourne, Madrid, Cape Town,
Singapore, São Paulo, Delhi, Mexico City

Cambridge University Press
The Edinburgh Building, Cambridge CB2 8RU, UK

Published in the United States of America by Cambridge University Press, New Yor

www.cambridge.org
Information on this title: www.cambridge.org/9781107642027

First published 1912
First paperback edition 2013

A catalogue record for this publication is available from the British Library

ISBN 978-1-107-64202-7 Paperback

PREFACE

THIS book needs little preface. In accordance with the prescribed scheme of this series of County Geographies the author has tried to describe the physical features of Oxfordshire, its people, its industries, history, antiquities, architecture and famous men, and to recall the story of Oxford with its seven centuries of University life, the Saxon and Norman stronghold that played a great part in the Danish strife, and in the civil wars of the twelfth and seventeenth centuries. It has been a pleasant task to wander again through the old-world towns and villages of the county, and to attempt to lead others to admire its remarkable churches, its fine examples of domestic and collegiate architecture, and to interest them in its history.

The author desires to express his thanks to all who have assisted him in the writing of this book, to Mr Llewelyn Treacher, F.G.S., whose aid has been most valuable in the geological and kindred sections, and especially to the General Editor, Dr Guillemard, for his editorial care.

P. H. D.

January, 1912.

CONTENTS

ILLUSTRATIONS

MAPS

The photographs on pp. 74 and 76 were kindly supplied by the Oxford University Press; thanks are also due to the Radcliffe Observer for his assistance in procuring the photograph on p. 63.

1. County and Shire. The Origin of Oxfordshire.

If we look at a map of England we see the whole country parcelled out into shires and counties, each of which has its own story to tell of its origin and formation. A remarkable difference exists with regard to their origin, as great a difference, indeed, as in their size and characteristics. Several were old kingdoms long before one king ruled over a united English land. The southern counties are much older than those further north. The name Kent preserves the memory of an old British tribe, the Cantii, who held the south-eastern corner of our island long before Julius Caesar came to try to conquer Britain. Other counties record Saxon kingdoms, such as Sussex, the region of the South Saxons; Essex, that of the Eastern Saxons; Middlesex that of the mid-Saxon kingdom; while the Angles held East Anglia and divided themselves into the North Folk, or Norfolk, and the South Folk, or Suffolk. The West Saxons were a powerful people and held Berkshire, Wiltshire, and Gloucestershire, and part of Somerset—the later Wessex. Wessex had its own folk-moot and its independent king. These districts were subsequently divided into the shires which we now see on the map.

On the north of Wessex was the powerful kingdom of Mercia, which maintained its integrity until the tenth century. It had many warrior rulers who waged war on the adjoining kingdoms, fighting incessantly with the neighbouring people of Wessex. About the year 912 the partition of this large stretch of country was made. It was separated into *shires*, i.e. *shares* or divisions, parts *shorn* off—for the root-word is the same in each case—these new shires all bearing the name of the chief town around which they were grouped. Thus, Staffordshire, the shire of Stafford, Warwickshire, Worcestershire, Leicestershire, Nottinghamshire, Northamptonshire, and others were founded, and among them our shire of Oxford sprang into being about the little town which was destined to become a great city, and on account of its university one of the most famous places in England.

It was not then so important a town as it afterwards became, but its position on the great river near the chief ford would render it a place of consequence. Oxford used to be written Oxenford, and it may be that the name is derived from these beasts of burden, "the ford of the oxen," though some learned men tell us that the first syllable comes from the Celtic word Ouse, meaning a river, and that the name means the ford across the river. Earthworks, cromlechs, camps, and roads, tell of the earliest people who inhabited the district. On the west they formed their tribal boundary where the Edge Hills overlook the plains of leafy Warwickshire. On the south the Thames formed a means of defence, and opposite the Berkshire stronghold at Sinodun, near Wallingford, stands the camp called the Dyke Hills, protected by a double vallum and a trench.

Long before the Romans came, a warlike Celtic tribe called the Dobuni dwelt in the district now called Oxfordshire. They were surrounded by other strong tribes, the Carnabii on the west, the Coritani on the north, the Atrebates on the south (whose chief city was Calleva Atribatum, afterwards the Roman Silchester), and the Catuvelauni on the east. There was much fighting between these tribes, and the Dobuni extended their sway to the Severn. Then they were harassed by their neighbours, the Catuvelauni, who inhabited the district now called Buckinghamshire and had at one time for their chief Cunobeline, of whom Shakespeare wrote, calling him Cymbeline.

When Julius Caesar came to Britain he could not penetrate the forests of the future Oxfordshire, and it was left to another great Roman leader, Aulus Plautius, to subdue the Dobuni. We can find several traces of Roman rule in Oxfordshire, though they are not so numerous as in many other counties. The Romans remained in possession until about 410 A.D., when they withdrew to their native land, and Britain was left defenceless. Then came the Saxons and Angles, and this part of our land fell an easy prey to the West Saxons. At the beginning of the seventh century it was part of Wessex, the kingdom of the West Saxons. Then from the north-east came another powerful Saxon host, the Mercians, who contended with the West Saxons who had advanced from the south. Fiercely did the battles rage, first one side being victorious and then the other. Penda, King of Mercia, who ruled from 626 to 633, was a mighty warrior, and extended his rule over Oxfordshire, and by treaty with Cwichelm, King of Wessex, made the Thames the boundary-line between the kingdoms. But this arrangement did not last long. For two centuries the rival kingdoms contended, first one gaining an advance and then the other, until at last Egbert, King of Wessex, prevailed in 827 and brought Mercia under his rule. It preserved, however, its geographical boundaries and organisation, being ruled over by an Ealdorman, until at the beginning of the tenth century Mercia was divided, as already stated, into shires, each shire taking its name from its chief town.

With the coming of the Normans the word county was introduced. They applied that word in order to identify the old English "shire" with their own *comitatus*, the district of a *comes* or count. Thus Oxfordshire, "the shire of Oxford," became also known as the county of Oxford. A reference to the Domesday Survey shows that like many of the other southern shires it was divided among smaller landlords or into smaller estates than the great lordships of the midlands and the north. Hence the owners were not so powerful as the barons who led the revolts against the Norman and Plantagenet kings.

It is curious to note that although England became one kingdom the shires or counties retained for centuries their own peculiarities and local customs. They had each their own manners and social traditions. Kent,

TEMPORE REGIS EDWARDI Reddebat OXENEFORD
p theloneo 7 gablo 7 omib alius csuetudinib p annu
regi qdem . xx . lib 7 vi . sextar mellis . Comiti u Algaro
x . lib . adiuncto molino que infra ciuitate habebat .
Quando rex ibat in expeditione : burgenses . xx . ibant
cu eo p omib aliis . uel . xx . lib dabant regi . ut omes essent liberi .

Modo redd OXENEFORD . lx . lib ad numeru de . xx in ora .
In ipsa uilla tam intra muru qua extra ht . cc . 7 xl iii . dom
reddent geld . 7 exceptis his st ibi qngente dom . xxii . min
ta uaste 7 destructe qd geld non poss reddere .
Rex ht . xx . mansiones murales que fuer Algari . T.R.E.
reddtes tc 7 m . xiiij . solid . ii . denar min .
7 una mansion ht redd . vi . den . ptin ad Sciptone . 7 alia de
iiij . den ptin ad Blochesha 7 tcia redd . xxx . den ptin ad Rise
berge . 7 ii . alias de . iiij . den ptin ad Luiforde in Buchingeha
scire . Una ex his . e uasta . 7 muru reficiunt .
Ideo autem uocant murales mansion . qa si opus fuerit 7 rex pceperit
Ad tras qs tenuit Albicus com ptin . i . eccla . 7 iii . mansiones .
haru . ii . iacent ad ecclam S. MARIE . reddtes . xx viij . den .
tcia iac ad Bureford redd . v . solid .
Ad tras qs W . com tenuit ptin . xx . mansion redd . vii . solid
Ex his st . iiij . uaste .
Archieps Cantuar ht . vii . mansion redd . xxx viii . den . Ex his
iiij . st uaste . Eps Wintoni . ix . mans redd . lxii . den . Ex his
iii . st uaste . Eps Baioc . xviij . mans redd . xiij . sol . 7 iiij . den .
Ex his . iii . st uaste . Eps Lincol ht . xxx . mans . reddtes
xviij . sol . 7 vi . den . Ex his . xvi . st uaste . Eps Constantiens
ht . ii . mans redd . xiiii . denar . Eps Hereford ht . iii . mans
redd . xiij . den . Ex his una uasta . e .

Portion of Domesday Book relating to Oxfordshire

for example, retained its custom of gavelkind, whereby the youngest son, and not the eldest as in other counties, was entitled to his sire's property. Certain peculiarities in the building of Wadham College show that it was erected by Somerset masons. Even in such things as shoeing horses each county had its own peculiar custom. Thus Charles II in his flight was once detected by his horse's shoes having been made in four different counties.

Since the population of the country has become more migratory, and railways quickly convey us from one end of the kingdom to the other, and the universal application of economic laws to the whole realm has come into force, these peculiarities of shires have for the most part disappeared.

2. General Characteristics. Position and Natural Conditions.

If we divide England into two fairly equal parts by a line running from east to west, Oxfordshire lies in the centre of the southern half. There is no coal in Oxfordshire and few openings for modern industrial activity. It is essentially an agricultural county. Camden well described Oxfordshire as "a fertile country and plentiful, the plains garnished with cornfields and meadows and the hills beset with woods."

In former days Oxfordshire was noted for its wool trade, and other enterprises which depended on the water-power of its rivers and streams. These rivers form a distinguishing feature of the county. If we take a map of the county and colour all the rivers blue, we see that it is one of the best watered shires in England, showing, indeed, a perfect network of watercourses. This was observed by Dr Plot, who produced in 1677 his *Natural History of Oxfordshire.* He wrote "that Oxfordshire is the best water'd county

Burford

in England, though I dare not with too much confidence assert, yet am induced to believe there are few better." He might have stated the fact with certainty, if he had known the county better, and we shall presently consider the great river and its tributaries, and perhaps enumerate some of the "three score and ten at least of an inferior rank, besides smaller brooks not worth mention."

These rivers had a great advantage for the trade of the shire in olden days, as they produced a large amount of water-power upon which the primitive wool trade and cloth-making largely depended. Five Oxfordshire market towns are all on rivers. These are Burford, Chipping Norton, Henley, Thame, and Witney, and as we have said, Oxford owes much to its position near the head-waters of the navigation of the Thames. But the days of water-power and waterways are past and gone. Steam engines have long since supplanted the old water-wheels, and in these days of railways we no longer depend on barges and rivers for the conveyance of our goods.

The river Thames greatly promoted the trade of the county in former days. It was the principal means of transit of goods and the great highway of traffic. As early as 1205 King John gave licence to William FitzAndrew to have one vessel to ply on the Thames between Oxford and London. Stone for the building of Eton College was conveyed from the Headington quarries, and before the advent of railways this important river was the great highway, and brought trade and prosperity to the counties along its banks. In ancient times it brought also sundry dangers and disadvantages, and troublesome visitors. The Danes sailed up the Thames and burned and pillaged and slaughtered along its banks. In 871 they came as far as Reading in Berkshire. Later they settled at Oxford. In the latter part of the tenth century they roamed about the country plundering and destroying. In 1006 they sacked and burnt Oxford, and in the next year marched again through Oxfordshire, but at length

Mapledurham Water Mill

under the rule of Canute in 1018 a council was held at Oxford when both Danes and Angles agreed to observe the laws of Edgar, and live in peace. In those days of rapine and slaughter the great river would not have been considered an advantage to Oxford. The bed of the Thames contains stores of weapons which conquerors and conquered have dropped from their warships and canoes, and when the river is dredged we often find stone implements, bronze weapons, swords, and daggers.

Another great natural feature of the shire was the abundance of wood and forest land. It must be remembered that, in its earliest days, England was very largely uncleared scrub or woodland, and in parts true forest land in the modern restricted sense of the term, i.e. covered with large trees. At one time Oxfordshire was almost covered with forests: it was practically a continuous woodland. The royal chase of Woodstock lay on the north, and contiguous was the forest of Wychwood (perhaps "the wood of the Hwicci"). Near Bicester was the forest of Bernwood. Stowood, Beckley, and Shotover lay on the east. The Chilterns on the south-east were covered with wild thickets and dense beech-woods, and Bagley Wood extended on the south-west. These forests, full of deer, attracted the attention of Norman kings, and became the favourite royal hunting grounds. In the fourteenth century the Chiltern Woods became the haunt of thieves and robbers, and a Steward was appointed to exterminate these pests and to guard travellers and protect the inhabitants from pillage and murder. Though the robbers have long since vanished, the Stewardship remains. Its salary is £1 per annum, and when a Member of Parliament wishes to resign his seat, he accepts the office of the Stewardship of the Chiltern Hundreds. During the great Civil War of the seventeenth century, there was a terrible destruction of timber. Camden tells how the rich and fertile county

Woodstock

with hills covered with great store of woods, had been shorn of its beauty, and Dr Plot says, "The hills before the late unhappy wars were well enough beset with woods, where now 'tis so scarcy that 'tis a common thing to see it sold by weight."

Another cause of the destruction of Oxfordshire woods

was the high price of corn during the French wars at the close of the eighteenth century, when hundreds of acres of woodland were grubbed up in order to produce wheat, which was sold at an enormous price. The farmers and squires had great difficulty in making the land fit for cultivation, and owing to the destruction of the trees the price of wood was nearly doubled.

We may conclude that the general characteristics of the county are agricultural; that its natural conditions are abundance of water and (until recently) of forests; that the great number of streams producing water-power provided, in old days, means for cloth-working; that the woods provided a great trade in timber; and that the Thames was the principal means of traffic connecting Oxford with Windsor and London, and thus bringing trade and commerce to the Oxfordshire city.

3. Size. Shape. Boundaries.

Oxfordshire cannot boast of its size. Thirty other English counties have a larger area. Compared with Yorkshire, Lincolnshire, Devon, or Norfolk, it seems a small county, but it contains 483,626 acres or 751 square miles. It has a curiously irregular shape. Its greatest length—measured from the most northern point, the Three Shire Stones, a mile north of Claydon to Caversham near Reading—is exactly 50 miles. Its greatest breadth is about 33 miles, measured from the Thames near Kelmscott to the Great Ouse near Finmere.

The Thames Valley near Goring

But near Oxford we can walk across the county in an hour and a half, as the distance is under seven miles, and though the width of the county increases a little as we proceed south, it never exceeds a dozen miles.

Fanciful geographers have compared the shape of the county to that of a lute. There are, however, some reasons for this eccentric shape. Its boundaries follow certain natural features which doubtless appeared good frontier lines and convenient demarcations to those who first mapped out Oxfordshire. Thus the great river Thames with its winding course made an admirable boundary on the south, separating it from Berkshire. The Chiltern Hills with their crowns of beeches form a good limit on the south-east, and on the north and north-west the grand range of the Edge Hills provides a frontier line separating the county from Warwickshire. No one can now tell the origin of the boundary-lines separating from Northamptonshire on the north, from parts of Buckinghamshire on the east, and from Gloucestershire and a small detached part of Worcestershire on the west and north-west. Probably these lines followed old tribal divisions.

The counties that surround Oxfordshire are Berkshire on the south, Buckinghamshire and Northamptonshire on the east, Warwickshire on the north and west, and a detached portion of Worcestershire, and Gloucestershire on the west.

We will trace the boundaries of the county, beginning our peregrinations with the Three Shire Stones on the extreme north. Thence the line dividing it from Northants proceeds in a south-easterly direction, the ground gradually rising from 365 feet above the sea-level to 455 feet, past Wardington to Chalcombe (Northants.), then turning sharply to the west to the Cherwell river, and then south, pursuing an irregular course following the boundary of the borough of Banbury. The line then follows the

The Three Shire Stones
(*Oxfordshire, Warwickshire, Northamptonshire*)

course of the Cherwell as far as Clifton. It then turns east by Aynho (Northants.), past Souldern to the Ouse, where it leaves Northamptonshire, and the Ouse separates the county from Buckinghamshire until Water Stratford is reached.

The boundary-line then turns south, following the

course of the Roman road which leads to Bicester. Skirting the grounds of Finmere, and leaving the Roman road at Finmere station, it pursues an irregular southern direction with several minor twists and turns past Godington. It cuts across the London and North-Western Railway line, crosses the Ray stream and the Roman Akeman Street, goes by Piddington to near Brill (Bucks.), and then turns west passing the camp on Muswell Hill. It then makes for Murcot and turns south, past Studley Priory and its woods, through a well-timbered country, Shabbington wood being on the Buckinghamshire and Waterperry wood on the Oxfordshire side. The line then bears east and south towards the Thame river which it follows east to Thame. A tributary of the Thame, the Ford Brook, now forms the boundary until we reach the manor house of Aston Sandford, and then the line skirts the double peninsulas of Kingsey and Towersey. Turning south again, it crosses the Icknield Way, and, a little farther, reaches the foot of the Chilterns near Bledlow Cross. Here it mounts to the ridge, and following a very irregular and zigzag course, descends on the farther side to Stonor Park, and passing Henley Park eventually reaches the Thames by Fawley Court.

The southern boundary of the county follows the course of the Thames river, and will be traced in the section which describes the rivers of the shire.

The western boundary leaves the Thames at Kelmscott (the abode of St Kenelm) and divides Oxfordshire from Gloucestershire. It proceeds in a north-westerly direction, keeping east of Lechlade, and for a short distance follows the course of the river Leach, branching off soon after it passes Little Faringdon towards

Four Shire Stone

Broughton Poggs. It crosses Akeman Street, and the road from Burford to Cirencester by Fourmile House afterwards forms the boundary-line for a short distance.

Crossing the Windrush a little to the east of Great and Little Barrington, it rises to a height of 607 feet by Downs Barn, and goes past Idbury Camp to Bledington (Gloucestershire) afterwards following the Evenlode for a short distance. The boundary then passes by Chastleton Camp and House, separating Oxfordshire from a detached portion of Worcestershire. A peninsula juts out here, towards Moreton-on-the-Marsh, extending to the Four Shire Stone. The ground rises near Chastleton to a height of 749 feet, and the line proceeds in a north-easterly direction to the Rollright villages and Hook Norton. The level now descends to 371 feet, then rises to the escarpment of the Edge Hills, the boundary just avoiding the beautiful house of Compton Wynyates in Warwickshire, from which county Oxfordshire is here separated. The line then pursues a zigzag course, turning south-east near Hornton, and a tongue of Warwickshire juts into Oxfordshire by Shotteswell. It then resumes its irregular northerly course to the Three Shire Stones, where we began our peregrinations.

In old maps of counties we often find detached portions of the shires entirely surrounded by another county. These have often much historical importance, such as the Hundred of West Meon, situated in Sussex, but belonging to Hants. This is a striking survival from the settlement of Jutish Meonwara (i.e. the men of Meon), who were absorbed by the West Saxons of Hants. Oxfordshire is no exception to this. In ancient maps Lillingstone Lovell, Boycott, and Leckhampstead, all in Buckinghamshire and entirely surrounded by that county, are recorded as being parts of Oxfordshire; just as a little bit of detached Gloucestershire, including the parish of Shenington, joins the county on the Warwickshire border. It is an interesting problem to ascertain how these bits of counties became detached. In some cases, no doubt, the separation took place in early times by conquest or agreement; but in most instances the detached portion belonged to some great landowner, whether a private individual or—as was not uncommon—an ecclesiastical body, whose chief estates were in that county, and therefore for manorial rights and assessment was permitted to consider the outlying property as part of the shire. The name Lillingstone Lovell points to this interpretation. The Lovells were a great Oxfordshire family. They gave their name to Minster Lovell in Oxfordshire, and Lillingstone Lovell was also their property, and so counted as part of Oxfordshire.

These detached portions of counties are now only a matter of historic interest. By an Act of Parliament passed in 1844, all such parts were annexed to the county by which they were surrounded, and these bits of Oxfordshire have long since been amalgamated with Buckinghamshire.

4. Surface and General Features.

The surface of our shire is pleasantly diversified. Nothing could be further from monotony. Within its limits we find rolling plains, often wolds or almost moors, swift shallow streams, bare upland, and wooded valleys.

On the south-east are the glorious Chilterns, rising to a height of over 800 feet, clothed and crowned with beechwoods. Shirburn Hill stands above the rest with a height of 827 feet, but Nettlebed (700 feet) with its woods and commons is the pride of the range. Camden, who wrote at the end of the sixteenth century, speaks of "the hills beset with woods; stored in every place, not only with corn and fruit, but also with all kinds of game for hound and hawk, and well watered with rivers plentiful." A delightful description truly! But time has wrought some changes. Many woods have disappeared, and hawks are no longer generally used for killing game, except for amusement by a few enthusiastic followers of the sport, though hounds still pursue the fox and the hare. The scenery, however, is much the same as when the old geographer visited the county, and the rivers pursue their course to the sea and still abound with fish.

Oxfordshire scenery varies much in different parts. We have rich meadows and overhanging woods by the sides of the rivers, as well as bleak and bare uplands with their stone walls and shelterless downs. The slopes of the Chilterns are still clad with beechwoods, and in the north we find enormous fields without a tree to cast a shadow. But there is also in the north some beautiful scenery, the country being diversified by lovely wooded valleys. In the southern part of the county, in the region of the Chilterns, the lover of nature finds at every season of the year much that pleases him. In the spring, the young green of the cornfields competes with the tender leaf of the stately beech-trees with which many of the slopes are covered. In summer we pass from great stretches of golden grain to the leafy shades of the woods, where the sun's rays flicker on the smooth straight trunks of the beeches and the soft brown bed of leaves beneath; while in autumn the wooded hills are a mass of colour, the bright scarlet of the wild cherry and maple, the green and yellow of the oak, and the rich old gold of the beech, lit up by the clear bright sunshine, blend into as brilliant and harmonious a picture as we can find in any part of England. Nor, in its way, is winter a less beautiful season of the year, when the pasture-covered hills, dotted with dark juniper shrubs and darker box and yew trees, alternate with the brown arable land, which fully displays the soft and graceful curves caused by the action of the weather for centuries on the chalk formation of the soil. The beechwoods have changed their green coverings for the rich purple of next year's buds, and each tree stands clear against the sky, its delicate tracery suggesting the lace-work which the deft fingers of the country-women contrive to weave so skilfully.

The multitude of rivers, of which we shall presently have to speak, constitute a marked feature of the county. The extent of the woods and forests was another noteworthy characteristic, but the area covered by timber has been greatly reduced. The bogs and wastes of Otmoor—the home of the wild fowl, until drained, reclaimed, and enclosed at the beginning of the last century—still constitute a peculiar feature of the county. The land is still swampy and flat, being situated on a broad belt of clay, and the Ray stream only partially drains it and conveys its waters into the Cherwell river. We can get a fine view over this somewhat desolate region from Arncott's wood at Beckley.

Small valleys are numerous in the northern plateau with streams flowing through each, the slopes covered with pasture. There are few woods. Westward of this there is a region of long gently-rising slopes swelling to rounded hills, the outlying spurs of the Cotswolds. Stone walls often take the place of hedges, and the country is cold and bleak save when the summer sun shines and it becomes very hot and dry. Proceeding southwards towards Burford there is a hilly district, bleak and monotonous, with few trees or hedges, the latter being replaced by stone walls. The grand old forest of Wychwood no longer provides a hunting ground for the kings of England, and little of its woodland remains except around Leafield and Cornbury. Between Burford and Witney the country is hilly with small valleys running down to the Thames. The banks of the upper Thames are very beautiful, with low-lying meadows, tall hedges, and few trees. If we follow the course of the river to Oxford we may notice there the famous Port Meadow and Christ Church meadow, the reed-girt Cherwell and the rising slopes of Headington (322 feet) and Shotover (562 feet).

A survey of the county discloses a large number of parks, of which there are no less than forty. These with their magnificent timber and their ornamental lakes constitute an important feature of Oxfordshire.

The ground varies very much in elevation. The

surface of the Thames at Henley is only 108 feet above sea-level. At Oxford it is 187 feet. The Cherwell is 302 feet above the sea at Banbury. Tadmarton Heath in the north is 653 feet, Pitch Hill 705 feet, and the summits of the Edge Hills vary from 600 to 700 feet in height. On the other side of the county there is Britwell

Christ Church Meadow

near Watlington with a height of 753 feet, Nuffield Common 673 feet, and Nettlebed and Shirburn which have already been mentioned. There are, therefore, considerable variations in the surface of the shire, and the lovers of picturesque scenery can find many beautiful districts within its boundaries.

5. Rivers.

The rivers, as already stated, form a distinguishing natural feature of the county. They are so numerous and are fed by so many streams and brooks that almost the whole of the shire is covered with a network of water-courses, with the exception of the two plateaus at the extreme north and south of the county. What Dr Plot said with hesitation concerning our Oxfordshire streams we may assert with confidence. No county surpasses Oxfordshire in regard to the number and abundance of its rivers. The great river Thames flows along its whole southern boundary, and it has besides the Cherwell, Evenlode, Windrush, and Thame. All these flow into the Thames and have also their own tributaries. Thus, among many others, the Glyme and the Dorne add to the waters of the Evenlode, and the Ray to the Cherwell.

Dr Plot tells us that these rivers are "of so quick a stream, free from stagnation, and so clear...that few (if any) vappid and stinking exhalations can ascend from them to corrupt the air. And so for standing Pools, Marish or Boggy grounds, the parents of Ague, Cough, Catarrh, they are fewest here of any place to be found." Perhaps in this the learned doctor erred, or perhaps rivers like human beings sometimes change their nature. The Thames is swift enough, especially in times of flood, and was far more swift in Dr Plot's day, before it was bridled with locks; but the Thame, Evenlode, and Cherwell are somewhat sluggish streams and have many weeds, which are not seen in swift currents.

The conditions have changed since very early times. The bed of the Thames was once far less deep than it is at present, and the river extended itself on each side. Moreover the hills and vales were covered with dense forests, and these would tend to increase the rainfall and to make the atmosphere humid. The Thames is navigable for large barges and small steamers as far as Oxford, but it does not, of course, become a tidal river until long after it has passed the boundaries of the shire. It is also navigable from Oxford to Lechlade for boats and barges.

The Thames rises near Cirencester and runs southward into Wiltshire, and after receiving the Churn from the north of Cirencester and proceeding easterly by Cricklade, it unites near Lechlade with the Coln from the north and the Cole from the south, and becomes navigable. The Leach, from which Lechlade takes its name, also adds its waters to the river, and forms for some distance the boundary of the county. From this point the Thames forms the southern limit of Oxfordshire, and thence flows eastward, inclining to the north through an uninteresting country. It passes under Radcot Bridge, where a battle was fought in 1387 between Robert de Vere, Earl of Oxford, and the insurgent Barons, when the Earl only saved his life by plunging into the river; under Tadpole Bridge and New Bridge, where in the Civil War a fight took place. Here the Windrush joins the Thames. At Bablock Hythe there is a ferry, and there are many locks during this course of the river. Rounding the woods of Wytham it receives the waters of the Evenlode, but before reaching Oxford it divides itself into various small channels as it traverses the meadows of Wytham, leaving Oxford on the left. These streams, however, soon unite, and the river turns round the city and glides beautifully through

College Barges and Eights, Oxford

the meads of Christ Church. The Cherwell joins the Thames where the College barges are moored. Proceeding still south-eastward past the old mill at Sandford, the Norman church of Iffley, Radley, and the lovely woods of Nuneham, it flows in a westerly bend to Abingdon (Berks.) where it receives the Ock. Turning south and then east, past Culham and Clifton Hampden, it reaches Dorchester by a semicircular course, where it is joined by the Thame, and then runs south-eastward to Wallingford (Berks.). Past North Stoke and South Stoke it glides, and then reaches its most lovely scenery. All

Whitchurch

the way from Goring to Henley, save for a small uninteresting reach at Reading, it is girt by beautiful woods on one side and not less beautiful meadows on the other. Whitchurch, Mapledurham, and Purley are all beautiful. Eyots or Eyes (e.g. Sonning Eye) clad with willows, add diversity and beauty to the scene, and by their name preserve the old Saxon word for island. Near Reading the Thames receives the waters of the Kennet and turns in a north-easterly direction towards Sonning and Shiplake, welcoming on its way the Loddon river, both these streams coming in from the south. Soon it reaches Henley, and

Henley Bridge

opposite Remenham leaves the county, proceeding towards Windsor, London, and the sea. The length of the course of the river from the point where it first touches Oxfordshire to Henley, where it quits the county, is about 70 miles. With its tributaries it drains about 5000 square miles of country.

The tributaries of the Thames that run through the

county have cut their way through the limestone hills and made narrow valleys.

The Windrush rises in Gloucestershire, and entering Oxfordshire not far from the ancient town of Burford, in the delightful region of the Cotswolds, flows through the county a distance of about 16 miles. It is a land of breezy downs and bare hill slopes with old grey farm buildings dotted here and there over the fields. The Windrush runs clear and swift between pleasant meadows that go sharply up to the north, where stand the last trees of the old forest of Wychwood. It is unlike the rest of the Oxfordshire rivers with their muddy banks and sluggish currents. Before reaching Burford we see on the right the picturesque garden and farmhouse at Upton. Upton mills, once turned by the river, belonged to the Earl of Warwick, the King-maker. On the left is Taynton with its interesting Decorated church, which has a good Early English chancel. Taynton was the home of the Harmans, who afterwards removed to Burford Priory when it was dissolved. Burford town (the Borough-ford), with all its charms of history and romance, we shall visit again, and the Windrush pursues its course till it comes to another ford, called Widford (doubtless Wide-ford), a desolate place, an extinct parish with a deserted church dedicated to St Oswald, surrounded by an overgrown melancholy churchyard on the green banks of the stream. In vain its sweet bell in the bell-niche bears the inscription:

"Come ye all
At my call
Serve God all."

Half a century has passed since a few village folk obeyed that call, and the deserted shrine is fast falling into ruin. On its site once stood a Roman villa. A little farther down the stream we come to a river-side inn, near a bridge that spans the Windrush at Swinbrook, where formerly stood the splendid mansion of the powerful Fettiplace family, but both house and race have vanished. Of this family the old rhyme says :—

> "The Tracys, the Lacys, and the Fettiplaces
> Own all the manors, the parks, and the chases."

But now nothing remains but their monuments and brasses in the church. A little farther down the stream on the right bank is Asthall, where there is another bridge. Close by, the Akeman Street crossed the river by a ford. About a mile south is Asthall barrow, a relic of prehistoric times. The Windrush proceeds eastward, and passing the lonely Minster Lovell, which we shall visit again, soon reaches Witney.

The waters of the Windrush have certain peculiar properties which favour clothmaking. Dr Plot describes them as "abstersive," whatever that may mean. They are found, however, to be useful in blanket-making, and have brought prosperity to the little town of Witney. Fish, too, thrive on the "abstersive" quality of the Windrush waters and are vastly superior, we are told, to those in most other streams. From Witney the river turns south, passing through a tract of flat narrow land intersected by water-courses, known as "Ducklington Ditches," one of which is called Emm's Ditch, marking the boundary of Queen Emma's manor, the queen of King Ethelred and Canute. The village of Ducklington stands close to the river, with its fine church showing mainly Early English work, with examples of Norman, Decorated, and Perpendicular styles. Past the woods of Cokethorpe Park the Windrush flows, and then wends its zigzag course to Standlake and its British village, soon finding its way to the "stripling Thames" at Newbridge, which is nowadays not new at all, but has stood for centuries grim and grey with its curious triangular buttresses. On it a skirmish was fought on May 27, 1644, the day after Essex had occupied Abingdon. Between Witney and Standlake the Windrush flows as a double stream, the branches being from a quarter to half a mile apart.

The Evenlode rises in Gloucestershire and runs during part of its course almost parallel with the Windrush, entering the county at Bledington. Several tributaries unite their waters with the river, which flows past Chipping Norton junction—now called Kingham—in a south-easterly direction to Shipton-under-Wychwood with its graceful spire, and then turns north-east to Ascott-under-Wychwood past the old gabled Elizabethan manor house of the Lacys, called Pudlicote, curving round to Charlbury, a quiet little village on its banks. It bounds the old park of Cornbury, which has been a park since the year 1312–3, and the slopes of the valley are beautifully wooded. Hitherto the valley through which the river runs has been wide and open but it now becomes narrower, and the river continues in a south-easterly

On the Cherwell

direction till the woods of Wilcote run down to its edge. Its course becomes tortuous and soon the Blenheim woods appear on the left, with the tributary stream the Glyme, which rises near Chipping Norton and flows past Kiddington and Glympton and through the ornamental waters of Blenheim Park. The Evenlode then flows almost due south, joining the Thames opposite the woods of Wytham.

The Cherwell is the most important of the Oxfordshire rivers after the Thames, and has a course of thirty miles in the county. It rises in Northamptonshire, and on entering Oxfordshire near Wardington in the north of the county receives many tributary brooks and streams, flowing almost due south save for a slight curve after passing Hampton Gay. Its vale is wide and open during the upper part of its course, until it reaches Banbury, where it scoops out for itself a narrow valley similar to many other Oxfordshire rivers, forming very pleasant and attractive features of the scenery of the shire. For some distance it forms the boundary of the county until it passes near Clifton, having received the waters of the Sorbrook and Swere tributaries. The Swale, a small river from which Swalcliffe takes its name, is a tributary of the Sorbrook. Passing the well-wooded park of North Aston, the Cherwell creates for itself one of its fairest reaches until Heyford bridge appears. The old name of the village was Heyford-ad-Pontem, a very early bridge having been built here by Robert D'Oilly, lord of Wallingford in the Conqueror's time. Its name signifies the presence of a ford that preceded the bridge.

Other bridges span the stream at Northbrook and Enslow, and the Akeman Street crosses the river between them where there was once a ford. The river now passes by a scene of desolation, the decayed hamlet of Hampton Gay. The village has only three houses, and it is a melancholy sight to gaze upon the ruins of the manor house recently destroyed by fire, and on the burnt paper-mills. The banks are very steep here, and opposite stands the picturesque church of Shipton-on-Cherwell. When we have passed the lofty spire of Kidlington, we see the Ray river, which, after passing through Bicester and Otmoor, here comes in on the left bank. The Cherwell now flows on past the beautiful manor house of Water Eaton, past the magnificent pile of Magdalen College, and pours into the Thames just below the College barges. Its beautiful lower reach, bedecked with lilies and lovely water plants, is much frequented in summer by such undergraduates as prefer dreaming in a punt to the more strenuous work of the Eights.

The last Thames tributary that runs through the county is the Thame, which rises in Buckinghamshire and flows through the town that takes its name from the river. It is not a very interesting stream, pursuing its uneventful course for the most part through flat meadow-land. A bridge spans it at Chiselhampton, long and narrow, with bold projecting cut-waters, carrying the road over two branches of the stream and low-lying meadows. Here, on the morning of the battle of Chalgrove Field, a skirmish took place between the royal forces under Prince Rupert and the army of the Parliamentarians led by John Hampden, who for some time kept the Prince at bay. On the other side of the river is Stadhampton, where John Owen, Cromwell's chaplain, was born. Past Newington and Drayton the Thame flows, and just below Dorchester reaches the place

> "Where beauteous Isis and her husband Thame
> With mingled waves for ever flow the same."

South of this there are no streams of any importance, the Chiltern district being badly supplied with water except near the Thames.

The Stour rises in the county near Tadmarton heath and Hook Norton and flows westward through Swalcliffe common, soon leaving the county. On the east side of the county the river Ouse touches it, forming for some distance the boundary of the shire.

6. Geology and Soils.

Geology deals with the rocks which form the outer part or crust of the earth, their structure, contents, and relative positions. It further investigates their mode of formation and the laws which regulate their arrangement, and describes the influence they have on the scenery, the agriculture, and the mineral wealth of the country.

Rocks are usually grouped in two classes, the Igneous and the Aqueous. These, as their names imply, owe their condition, the former to the action of fire, or more correctly the internal heat of the earth, and the latter to the action of water. As the Igneous rocks do not occur within the boundaries of our county and have only a remote connection with its geology they need not be further noticed here.

The Aqueous rocks are generally found arranged in layers or strata, and were originally laid down beneath the water in lakes, estuaries, or seas. The constant movement of the water wears away the surrounding land. Waves dashing against the cliffs break off fragments which, falling on the beach, are worn into pebbles, sand, or mud. Rivers also bring down great quantities of sediment, the result of the waste of the land surface under the influence of frost and rain. These materials get washed away by tidal and other currents and deposited in beds on the sea-floor, the coarser near land and the finer farther off. Mixed with them are the remains of plants and animals which lived and died during the time each particular bed was being laid down.

In process of time, by local movements of the earth's crust, these beds perhaps become elevated into dry land, the water moves off into other depressions in the earth's surface, and the same process is again gone through. Thus records of the former distribution of land and sea, of the forms of life which then existed, and consequently of the climates which prevailed, are preserved in the rocks.

The crust movements have affected the rocks in different degrees. Some while being elevated were sharply folded, and had their higher parts worn away before they were again depressed and another stratum laid down upon them. Others retain almost their original horizontal position so that the newer beds only gently

	Names of Systems	Subdivisions	Characters of Rocks
TERTIARY	Recent Pleistocene	Metal Age Deposits Neolithic ,, Palaeolithic ,, Glacial ,,	Superficial Deposits
	Pliocene	Cromer Series Weybourne Crag Chillesford and Norwich Crags Red and Walton Crags Coralline Crag	Sands chiefly
	Miocene	Absent from Britain	
	Eocene	Fluviomarine Beds of Hampshire Bagshot Beds London Clay Oldhaven Beds, Woolwich and Reading [Groups Thanet Sands	Clays and Sands chiefly
SECONDARY	Cretaceous	Chalk Upper Greensand and Gault Lower Greensand Weald Clay Hastings Sands	Chalk at top Sandstones, Mud and Clays below
	Jurassic	Purbeck Beds Portland Beds Kimmeridge Clay Corallian Beds Oxford Clay and Kellaways Rock Cornbrash Forest Marble Great Oolite with Stonesfield Slate Inferior Oolite Lias—Upper, Middle, and Lower	Shales, Sandstones and Oolitic Limestones
	Triassic	Rhaetic Keuper Marls Keuper Sandstone Upper Bunter Sandstone Bunter Pebble Beds Lower Bunter Sandstone	Red Sandstones and Marls, Gypsum and Salt
PRIMARY	Permian	Magnesian Limestone and Sandstone Marl Slate Lower Permian Sandstone	Red Sandstones and Magnesian Limestone
	Carboniferous	Coal Measures Millstone Grit Mountain Limestone Basal Carboniferous Rocks	Sandstones, Shales and Coals at top Sandstones in middle Limestone and Shales below
	Devonian	Upper Mid Lower Devonian and Old Red Sandstone	Red Sandstones, Shales, Slates and Limestones
	Silurian	Ludlow Beds Wenlock Beds Llandovery Beds	Sandstones, Shales and Thin Limestones
	Ordovician	Caradoc Beds Llandeilo Beds Arenig Beds	Shales, Slates, Sandstones and Thin Limestones
	Cambrian	Tremadoc Slates Lingula Flags Menevian Beds Harlech Grits and Llanberis Slates	Slates and Sandstones
	Pre-Cambrian	No definite classification yet made	Sandstones, Slates and Volcanic Rocks

overlap them. Others again have been buried deep down, almost beyond the limits of observation.

Although a great variety of rocks exists even in a small country like England, it must not be supposed that more than a few can be found in any one locality. Some of the beds were originally of only small extent and others have been widely destroyed during periods of elevation, causing gaps to occur in the regular order of succession. However, when the whole country is explored, many of these gaps can be filled up by examples taken from other localities and it becomes possible to construct a table like that given on page 37, showing the position and age of any bed or formation relatively to the others. And these relative positions are constant. Once identify a bed as belonging to one of the great systems and we know that it is newer than those lower down the diagram and older than those above, although none of them may be seen for many miles away. Fossils are excellent helps in identifying strata, as each age had its own peculiar forms of life, the remains of which are preserved in the rocks of that period.

With most people the study of geology begins with an interest taken in fossils. These "medals of creation," as they have been called, so much like, and yet somehow unlike, the remains of creatures now living, and their position in the solid rocks far from the seas in which these organisms now live, early attracted the attention of the curious, and their beauty the desire of those who made collections. The small stone quarries which formerly abounded throughout the northern and central parts of

Oxfordshire were famous among the happy hunting-grounds of collectors like Plot (1677), Llwyd (1699), and Parkinson (1804) who described and illustrated the fossils of the county in books which now read like the narratives of early explorers of an unknown continent.

Although the geological exploration of Oxfordshire has long been completed and the rough outlines of its rock-history described, there is still work to be done if only in following the footsteps of the old masters through quarries, pits, and river valleys, discovering fresh details and seeing new aspects of the science, so that there ought not to be any lack of interest in geology as an intellectual recreation as well as a study of practical importance.

In Oxfordshire are found representatives of nearly every member of the Jurassic and Cretaceous systems together with traces of many of the later rocks. The oldest strata are exposed at the northern end of the county and the newer ones succeed one after the other as we proceed southwards. Generally each has a gentle dip to the south-east, in which direction most of them pass beyond the border, whilst north-westward they are cut off by denudation, the harder limestones forming steep escarpments and the softer clays low plains or valleys between.

Although strata belonging to the Triassic system were proved at a depth of nearly 700 feet from the surface in a boring made near Burford, with Carboniferous rocks 500 feet deeper still, the oldest formation exposed within the county is the Lias with its three divisions of Lower, Middle and Upper, each having its own distinctive features.

The Lower Lias consists of blue clay with many hard layers of impure limestone, and forms the low ground north of Banbury, extending thence over the broad plain at the foot of the Cotswold Hills. The Middle Lias is more sandy and contains a hard rock-bed known as the Marlstone, which stands up in the bold escarpment of Edge Hill and dips gently thence along the hill-tops near Banbury till it disappears beneath the clays of the upper division about Deddington. In places it contains iron ore, which has been worked at Fawley, Adderbury, and Hook Norton.

The Lias was deposited in a shallow sea bordered by land consisting largely of rocks of Carboniferous age, whose dark shales supplied the muddy sediment of which the strata are composed.

The sea swarmed with living creatures, chief among which were the great reptiles *Ichthyosaurus* and *Plesiosaurus* together with countless ammonites and belemnites. Remains of these may easily be found among the fossils from any Lias quarry.

The Lias is 1360 feet thick at Chipping Camden in Gloucestershire, but decreases to 627 feet in the Burford boring and to probably less in the southern part of Oxfordshire, the Upper and Middle divisions thinning very rapidly in that direction.

Next above it comes the great series of rocks known as the Oolites, from the fact that many of these limestones are composed of small round grains like the roe of a fish. The lowest member of the series, the Inferior Oolite, crosses the county in a narrow band near Chipping

Norton and consists of variable layers of rubbly oolite, marl, and ragstone. Its average thickness in Oxfordshire is about 30 feet, but it becomes thicker both north-eastward and south-westward, attaining in the latter direction a thickness of 250 feet in the Cotswold Hills, where it contains many divisions of strata with marine fossils. North-eastward in Northamptonshire its fossils are of an estuarine character, showing that a large river entered the sea in that area. A common fossil in Oxfordshire is the echinoderm *Clypeus ploti*, which it is said was once used for pound weights in the country districts. Now it may be occasionally seen forming the borders of flower-beds and other ornamental garden work.

The next formation is the Great Oolite, a series of massive limestones with interbedded marl bands. About 130 feet thick in Oxfordshire, it covers the high land or "wolds" in the centre of the county and has a steep escarpment facing the north-west and also along some of the river valleys. In the lower part of the series occurs the famous Stonesfield slate, a thin-bedded shelly limestone which has been worked since Roman times for roofing material, but is now very little used. Houses roofed with this material have an appearance which for picturesqueness will compare with any other kind of building. The fossils of the Stonesfield slate are of peculiar interest, especially the remains of small mammals which were recognised by the celebrated palaeontologist Cuvier as long ago as 1818. The latter are however extremely rare, but other fossils are fairly common and are saved by the quarrymen for sale to visitors.

A fine section across the Inferior and Great Oolite of Oxfordshire was recently exposed in the cuttings for the new railway from Aynho to Bicester.

Down the dip south-eastward the Great Oolite is overlaid by two thin but interesting formations, the Forest Marble and the Cornbrash. The former owes its name to the fact that formerly it was quarried in the old forest district of Wychwood, north-east of Burford, and the harder beds polished and used locally for chimney-pieces in the farm-houses. The Forest Marble consists of shelly limestones, frequently false-bedded, the result of having been laid down in shallow water with rapid currents. Its thickness in Oxfordshire varies from 15 to 50 feet. It abounds in shells of small oysters, and echinoderms are by no means rare.

The Forest Marble is succeeded by the Cornbrash, an old agricultural term applied to stony and brashy soils suitable for the growth of corn. The Cornbrash formation consists of irregular layers of rubbly limestone with occasional seams of marl and clay. Although only from 10 to 20 feet thick, it is remarkable for its uniform development throughout the county. Fossils, chiefly shells, are fairly common everywhere.

The members of the Jurassic system already noticed appear to have accumulated as shoals and sand-banks in a shallow sea not far from land. The abundance of organic remains, and above all the presence of corals, seem to indicate a warm climate. With the deposition of the Cornbrash we see signs of a coming change. That formation was laid down in deeper and more tranquil waters than the preceding beds, and at its close there was an irruption of mud into the sea and this continued till a thickness of 450 feet of clay was laid down over Oxfordshire. This formation is known as the Oxford Clay and is largely worked for brickmaking in the neighbourhood of the city. It is generally full of fossils such as ammonites and belemnites, while bones of huge reptiles are commonly met with. It was probably deposited with comparative rapidity, as there is no great change in the forms of life throughout its mass.

Presently the water became clear again and shallower, and sand-banks were formed in places. On these banks shells collected and gradually formed a solid foundation on which grew coral reefs. This is the Corallian or Coralline Oolite, a so-called episodal formation ; that is, it was laid down during an interval between two great clays and was only developed in patches instead of being wide-spread. It is found however at intervals along a line reaching from Weymouth to Yorkshire. In Oxfordshire it has been much worked in the quarries about Headington and Shotover. It contains many interesting fossils, chiefly echinoderms and corals.

At the close of this episode the muddy conditions returned and the Kimmeridge Clay was deposited. This formation, named after the village of Kimmeridge on the Dorset coast, has a thickness of 1200 feet in the south of England, but this dwindles to 100 feet in Oxfordshire. It is, as a rule, darker and more shaly than the Oxford Clay and may be seen in the brickyards on the western side of Shotover Hill.

At the close of the Kimmeridge Clay period the sea began to retreat from Oxfordshire, but before this finally took place sands and limestones of Portland age were laid

Diplopodia versipora, Corallian

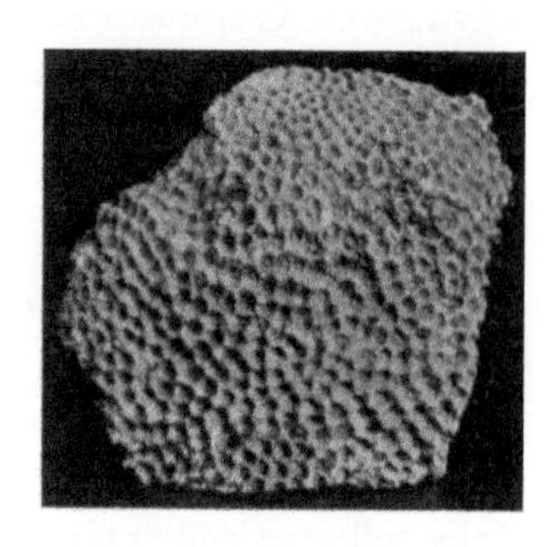

Isastraea explanata, Corallian

Ostraea deltoidea, Kim. and Cor.

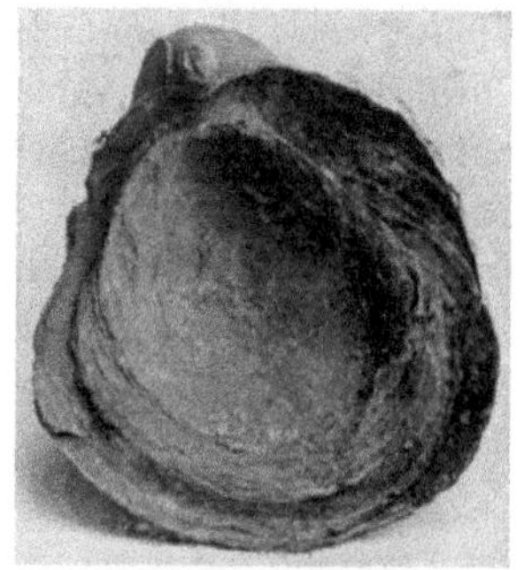

Gryphaea dilatata, Cor. and Oxf. Clay

Fossils of the Kimmeridge Clay and Corallian

down in the shallowing water. These now occur on Shotover Hill and about Garsington and Cuddesdon. Strata which have been referred to the overlying Purbeck Beds exist in small patches at Garsington and Brill.

With these the deposition of the strata belonging to the Jurassic system closed and a period of local elevation followed, during which great masses of the beds, especially of the later ones, were swept away, leaving, in some cases, only fragments to tell of their former wide extension.

We now come to the Cretaceous system of formations. The ironsands with inter-bedded clay beds on the top of Shotover Hill and extending along the ridge towards Wheatley, together with similar beds on Brill Common just over the border, are supposed to be of fresh-water origin and of the age of Wealden Beds of the south-east of England. The ironsands of Nuneham, Culham, and Clifton Hampden are of Lower Greensand age, and were deposited along the shores of a sea whose waters lay to the south-east. These compose the Lower Cretaceous series, and they suffered great local denudation before the beds of the upper part of the system were laid down upon them.

The Upper Cretaceous beds begin with the Gault, a dark blue clay locally about 200 feet thick. It has been worked for brickmaking near Thame and again at Culham, where the clay pit on the banks of the Thames shows a very interesting section. The Gault at this place rests directly on the Kimmeridge Clay, all the intervening formations which were almost certainly deposited here having been denuded before the commencement of Upper Cretaceous times. Next above the Gault comes the Upper Greensand, represented in Oxfordshire mainly by a rock known as Malmstone, about 90 feet thick. This rock, which has an appearance somewhat resembling chalk, differing from that rock however in being largely siliceous instead of calcareous, forms a well-marked ridge overlooking the clay plain north of Watlington, and is exposed in pits on Clare Hill.

About 12 feet of Greensand succeeds and then begins

Malmstone Quarry, near Watlington

the great Chalk formation which forms the Chiltern Hills in the southern part of the county. The Chalk, locally about 650 feet in total thickness, is a deposit of an open sea which contained very little land-derived sediment, being almost entirely composed of organic remains either entire or in a state of comminution. It is divided into three parts. The Lower Chalk is characterised by the presence of fossil ammonites and the absence of flints. It is also more marly than either of the other two divisions, and stretches out some distance over the low ground at the foot of the great escarpment. Above it and forming most of the face of the escarpment is the Middle Chalk, in rather thick beds and with very few flints. Recognisable fossils are scarce in this division. The Middle Chalk is also exposed along the side of the Thames Valley below Henley, the upper division having been worn away there by the river. The junction of the two divisions is well shown in a pit near Medmenham in Buckinghamshire. Between them there is a band of very hard rock containing many fossils of kinds not found in beds either above or below, and known as the Chalk Rock.

The Upper Chalk is familiar from the numerous bands of flints it contains. It now covers nearly the whole of the Chiltern Hills. Southward and eastward of the county higher beds of chalk with few flints are found, but if these were ever deposited in Oxfordshire they were denuded previous to the laying down of the Eocene clays and sands. The latter, comprising the Reading Beds and the London Clay, only occur in a few small outliers at Nettlebed, Binfield Heath near Sonning, and a few other places high up in the Chilterns.

The London Clay was probably the last of the solid rocks to be deposited in Oxfordshire and at the close of its period the county was finally elevated into dry land. From then till the present day its surface has been subject to the erosive action of atmospheric agencies in the form of water and ice. These have worn away the softer clays into low-lying plains and river valleys, leaving the more resisting limestones, sands, and chalk standing as hills and table-lands.

The force of falling rain-drops and the alternate freezing and thawing of the surface in winter break up the rocks more rapidly than the running water can carry the debris away, and so we find what geologists call superficial deposits occurring over the greater part of the county. In some places, especially on the Oolitic limestone tracts, there is nothing more than the simple soil formed by the disintegration of the solid rock in place. In others, as on the Chiltern Hills, there is a covering sometimes of considerable thickness of clay-with-flints, the result of the breaking up of clay beds of Eocene age mixed with flints derived from the chalk and left behind when the calcareous part of that rock has been dissolved. In the larger valleys again there is gravel and loam which the streams have brought down and have not been able to carry farther. These often occur as terraces at various heights along the sides of the valleys. In Oxfordshire some of these terrace-gravels are noticeable as containing flint implements, the work of the earlier races of men who inhabited the country. Wolvercote, Turner's Court near Wallingford, Caversham, and Shiplake are well-known localities for these relics.

As might be expected from the variety of rocks occurring therein, the soils of Oxfordshire are of many different descriptions. As a rule the Jurassic area is well cultivated. The Marlstone of the Middle Lias and many of the Oolitic limestones furnish a soil well suited for corn-growing. The Oxford Clay where not covered with drift or rain-wash is often cold and stiff, and expensive to cultivate, and its area is mainly under grass. The great plain which crosses the county along the foot of the chalk escarpment and connects the Vale of White Horse on the west with the Vale of Aylesbury on the east, contains some of the best corn land in the kingdom. The tops of the Chiltern Hills with their covering of poor stony clays were formerly thickly wooded; and, though on account of the high price of corn they were at one time much broken up for tillage, the arable land is now being replaced by plantations and the district is becoming a residential rather than an agricultural one.

7. Natural History—Fauna and Flora.

The conditions of life and the character and climate of Britain were very different in early times from what they are to-day, and if we want to know the kinds of animals which existed in former times it is necessary to learn what geology has to tell us about the physical aspect of the country. There was a time when this country was connected with the continent of Europe. The English Channel and North Sea did not exist and were mere valleys with rivers running through them fed by many streams. Where the North Sea now rolls there was the great valley of the Rhine; and as there were no oceans to cross, animals wandered northwards and westwards as they pleased from southern lands and made their abode here. Hence the country possessed the same kinds of animals as inhabited western Europe. Many of these have become extinct, but we find their remains as fossils, embedded in various superficial deposits. Amongst them are the bones of the bear, reindeer, hyaena, two kinds of elephant, rhinoceros, hippopotamus, urus, bison, and red deer. The question arises, how did these creatures become extinct in Britain? Why did they desert our shores, or leave their bones in the caves and hills?

It is a little difficult to obtain an entirely satisfactory answer, though one manifest explanation accounts for a good deal. In many cases, no doubt, they were driven southward by the severe cold of the Glacial Period and by the time the climate had ameliorated the sea had worked its way up the English Channel, broken through the chalk cliffs near Dover, and met the waves of the North Sea which flowed over the old Rhine valley. Hence the animals from the continent could not return to re-occupy their abandoned territory, and we have fewer species than France or Belgium. Ireland became separated from England before a sufficient length of time had elapsed to permit of its becoming peopled with all of the fauna of the latter country, and hence shows fewer species than England just as England shows fewer species than the continent. The same causes which produced a diminution of animals in Great Britain and Ireland as compared with those of the continent, also led to the lessening of the number of species of wild plants.

There are three sorts of plants in every country:—

(1) *Native plants*, the aboriginal species, which have always lived there: (2) *Denizens*, which are now almost native, but have at some remote period been introduced, e.g. the common elm or the Scots fir: (3) *Colonists*, or plants or weeds that owe their occurrence to the operations of man, e.g. the red poppy.

The total number of species of plants native in the British Isles is about 1750 and that of colonists, denizens, and aliens about 250; but, of these, 144 are confined to the neighbourhood of the sea, 17 are confined to Ireland, and 20 to the Channel Isles, while 200 are plants of northern latitudes, or are not found so far south as Oxfordshire. There are, therefore, 1369 species which might possibly be found in Oxfordshire. Mr Druce, whose book on the flora of the county is indispensable to the student, records that there have been found in Oxfordshire 847 native plants, 49 denizens, and 43 colonists, making a total of 939. Hence there are no fewer than 430 species which have not been recorded. It is possible that some reader may discover some of these. As Mr Druce says, "by a botanist finality can never be attained, since only a small portion comparatively of the actual surface of the ground comes under his observation and then only for a short time."

Botany is closely associated with geology. Some kinds of soil are favourable to some plants, while other plants require quite a different home and nurture. The region of Wychwood Forest on the limestone plateau rears many beautiful flowers. The orchid tribe is well represented in the county. We have the bird's nets orchis, a singular plant, its stem, seeds, and flowers all being of a dingy brown hue and looking at first sight like a withered stem. It derives its name from the short thick fleshy entangled fibres of its roots, which remind us of the sticks used by some of our larger birds in the framework of their nests. It flowers in June in the dark beechwoods of the Chilterns and is known by botanists as the *Listera nidus-avis*, the name *Listera* being derived from that of the distinguished botanist, Dr Martin Lister. Another orchid is the pyramidal orchis (*O. pyramidalis*), a lovely plant that prefers chalky soils and grows among grass. It has a tint of rich crimson purple, and flowers in July. Another is the green man orchis (*Aceras anthropophora*) whose yellow and green flowers show some faint resemblance to the human figure. The bee orchis (*O. apifera*), one of the prettiest of our wild flowers of a chalky soil, resembles a large velvety bee. It blooms in June and July. One orchis (*Ophrys muscifera*) assumes the form of a fly, and other species such as the marsh, the spotted, the early purple, and the green-winged orchis are to be met with.

North-west of Oxford, in the region of the old forest of Wychwood, is a rich and varied flora, amongst which are the hound's tongue (*Cynoglossum officinale*), the deadly nightshade (*Atropa belladonna*), lilies of the valley, the *Helleborus fœtidus*, lady's mantle, tooth-wort, and moonwort. By the pools are found many beautiful sedges, bulrushes, and mosses.

In the marshy meadows adjoining the Thames and Cherwell quite a different vegetation is found and the banks of the rivers abound with flowers, reeds, and rushes. A very characteristic flower is the fritillary (*Fritillaria meleagris*) commonly known as the snake's-head, very rare in many parts of the country but quite common here. Rare water-plants sometimes occur in this district, amongst them *Nitella mucronata*, only three times previously recorded in the British Isles. The fringed waterlily and the water violet are found in the river near Binsey, and the birth-wort (*Aristolochia clematitis*) at Godstow. The districts about Boar's Hill, west of Oxford, and about Shotover Hill, on the south-east, are the most interesting portions of botanising country in the centre of England. A long list might be given of the numerous flowers and plants that may be found there. The chalk hills of the Chilterns grow many plants which are not found elsewhere, and are gay with foxgloves, traveller's joy, gentians, and several kinds of orchids.

The fauna of Oxfordshire includes some interesting creatures. Foxes abound in the districts hunted by the South Oxfordshire, Heythrop, and Bicester hunts. The polecat is still found in the great woods, and the stoat loves the banks of the Thames, where he catches water-rats, young moorhens, and other small mammals and birds. The badger exists fairly numerously and otters frequent the Thames and its tributaries, and are often hunted by the Bucks otter hounds. The squirrel, dormouse, harvest mouse—the smallest of all rodents—wood or long-tailed field mouse, and house mouse are plentiful, though the last little pest is not so plentiful as it was. The brown or common rat has somewhat decreased in numbers. The water rat loves the Thames and its tributaries, and the bank vole or field vole is not uncommon. Hares abound in the county and the Peppard farmers hunt them in the Chiltern country near Nettlebed. They are especially plentiful near Thame, Churchill, Lyneham, Sarsden, and Chadlington. Rabbits have diminished in number in some places since the Ground Game Act was passed, but swarm on the juniper-covered downs, where they can do little harm. In some parts of the county both hares and rabbits are so plentiful that it would seem that the only effect of the Ground Game Act has been to lead farmers to preserve these animals more carefully. Deer are kept in twelve parks, all of them fallow deer. Before the enclosing of the forests there were plenty of red deer and many poachers and deer-stealers. "Burford bait" was renowned—an apple with a hook concealed in it, by means of which deer were caught and then killed.

The Thames is a grand river for fish. It was once a salmon river, before the numerous obstructions and the polluted mouth of the river prevented this fish from coming up from the sea in order to lay its spawn. The salmon will overcome all obstacles in order to beget its young, but it cannot face the polluted waters of the Thames at London. In recent years thousands of salmon smelts have been turned into the river, so that perhaps it may become a salmon river again.

The Thames trout is finer than any trout found in the other rivers of the British Isles, and from earliest times the Thames has been regarded as one of the most important angling rivers in England. Dr Plot states that "the plenty and goodness of the fish are a sure indication of the wholesomeness of the waters"; and that in 1674 the mayor and bailiffs of the city "between St Swithin's wear and Woolvercot Bridge," a distance of three miles, in two days caught 1500 jacks, besides other fish. Half a century ago many professional fishermen made a livelihood by netting the fish. Some people say that there are now fewer fish in the river than formerly, but thousands of fish have been placed in the waters—trout, bream, roach, perch, tench, rudd, and carp. Some of the fish taken are of great size. At Bablock Hythe in 1896 a pike of 26 lbs. was captured, and we hear of barbel weighing close on 12 lbs., and chub of 7 lbs.

The Windrush and Evenlode trout used to be famous, but the pollution of the waters has diminished their numbers. Chub and dace frequent the Windrush. Fishermen usually wage war on the coarse fish, such as pike and perch, and net and destroy them, in order to improve the condition of the trout. The natural condition of the rivers is much altered by man's agency, and into the Cherwell river have been introduced bream from Lincolnshire and perch from Windermere. The rudd seems to have disappeared from the Cherwell.

Dr Plot mentions some peculiar fish, and wonders also at two salmon taken at Lillingstone Lovell, about a yard in length, in a small brook (a branch of the Ouse) that a man may step over, little less, as the river runs, than two hundred miles from the sea. Crayfish and other freshwater shell-fish the learned Doctor discovered, some of which he searched for pearls, but lost his labour, as he could only find the smooth sort and not those with craggy rough outsides in which the precious gem (according to Sir Hugh Plat's work on *The Jewel-house of Art and Nature*) could alone be discovered. With such researches did this early naturalist and observer amuse and instruct his readers in the year 1677.

As we have already noticed there is a great variety in the soils and scenery of the county, some parts being bleak and monotonous, others abounding in woods, while along the rivers are wet and low-lying lands, where in former days the wild-fowl shooter earned a substantial livelihood. This variety of scenery has produced a corresponding variety in bird-life. The inclosure of wild tracts of heather-land and the tillage of the soil, the cutting down of woods and forests, have had their influence on ornithology, and it is not surprising that many birds have left their old haunts. During the last hundred years the kite, buzzard, harrier, raven, and bittern have practically disappeared, though they were fairly plentiful formerly.

Oxfordshire is too far inland to attract many of the winter visitors which are seen on the coast of East Anglia and Kent, but it has a large number of the species which are common in the south-eastern district of Britain. Gulls of various kinds come up the Thames during severe weather at sea, and find a pleasant retreat on the unenclosed Port Meadow north of Oxford. The numerous rivers, favoured haunts of the sedge and reed warblers, attract many migrants in their flight to and from their northern home, and many kinds of wild duck, swans, and other waterfowl.

Altogether there are 242 species of birds in the county apart from a few others, the presence of which has not been indisputably established. This number is larger than that of the birds in the neighbouring counties. Of this large list 60 are residents, 71 periodical migrants, and 111 occasional or accidental visitors. The Upper Thames was formerly a grand region for waterfowl of all kinds, but drainage and inclosures have diminished their numbers. Otmoor before it was drained and inclosed used to swarm with wildfowl, and in spite of all that has been done to spoil their paradise, many still remain, and thousands are captured in a decoy just over the Buckinghamshire border. The Chilterns with their juniper bushes attract the stone-chat, whinchat, and wheatear, and here too may sometimes be seen the dotterel and the stone curlew. The beautiful parks of Oxfordshire form excellent sanctuaries for birds, and waterfowl abound on the large ornamental lakes.

Oxfordshire people have invented some curious names for the birds that visit them. Thus a quail is called a twit-me-dick, a golden plover is a whistler, a hooded crow a dun crow, a missel-thrush a gizer. A green woodpecker is a hickle, whereas our Berkshire neighbours call the bird a yaffle or yaffler, and the humble hedge sparrow is surnamed Billy.

Those who wish to know more of the birds of the county should consult Mr O. V. Aplin's book on the *Birds of Oxfordshire*. Herein are "no stories told you of what is to be seen at the other end of the world, but of things at home in your own native country, at your own door, easily examinable with little travel, less cost, and very little hazard," as Master Childrey observed concerning another book as long ago as 1661.

8. Climate.

The climate of a country or district is, briefly, the average weather of that country or district, and it depends upon various factors, all mutually interacting; upon the latitude, the temperature, the direction and strength of the winds, the rainfall, the character of the soil, and the proximity of the district to the sea.

The differences in the climates of the world depend mainly upon latitude, but a scarcely less important factor is proximity to the sea. Along any great climatic zone there will be found variations in proportion to this proximity, the extremes being "continental" climates in the centres of continents far from the oceans, and "insular" climates in small tracts surrounded by sea. Continental climates show great differences in seasonal temperatures, the winters tending to be unusually cold and the summers unusually warm, while the climate of insular tracts is characterised by equableness and also by greater dampness. Great Britain possesses, by reason of its position, a temperate insular climate, but its average annual temperature is much higher than could be expected from its latitude. The prevalent south-westerly winds cause a movement of the surface-waters of the Atlantic towards our shores, and this warm-water current, which we know as the Gulf Stream, is one of the chief causes of the mildness of our winters.

Most of our weather comes to us from the Atlantic. It would be impossible here within the limits of a short chapter to discuss fully the causes which affect or control weather changes. It must suffice to say that the conditions are in the main either cyclonic or anticyclonic, which terms may be best explained, perhaps, by comparing the air currents to a stream of water. In a stream a chain of eddies may often be seen fringing the more steadily-moving central water. Regarding the general north-easterly moving air from the Atlantic as such a stream, a chain of eddies may be developed in a belt parallel with its general direction. This belt of eddies or cyclones, as they are termed, tends to shift its position, sometimes passing over our islands, sometimes to the north or south of them, and it is to this shifting that most of our weather changes are due. Cyclonic conditions are associated with a greater or less amount of atmospheric disturbance; anticyclonic with calms.

The prevalent Atlantic winds largely affect our island in another way, namely in its rainfall. The air, heavily laden with moisture from its passage over the ocean, meets with elevated land-tracts directly it reaches our shores—the moorland of Devon and Cornwall, the Welsh mountains, or the fells of Cumberland and Westmorland—and blowing up the rising land-surface, parts with this moisture as rain. To how great an extent this occurs is best seen by reference to the map of the annual

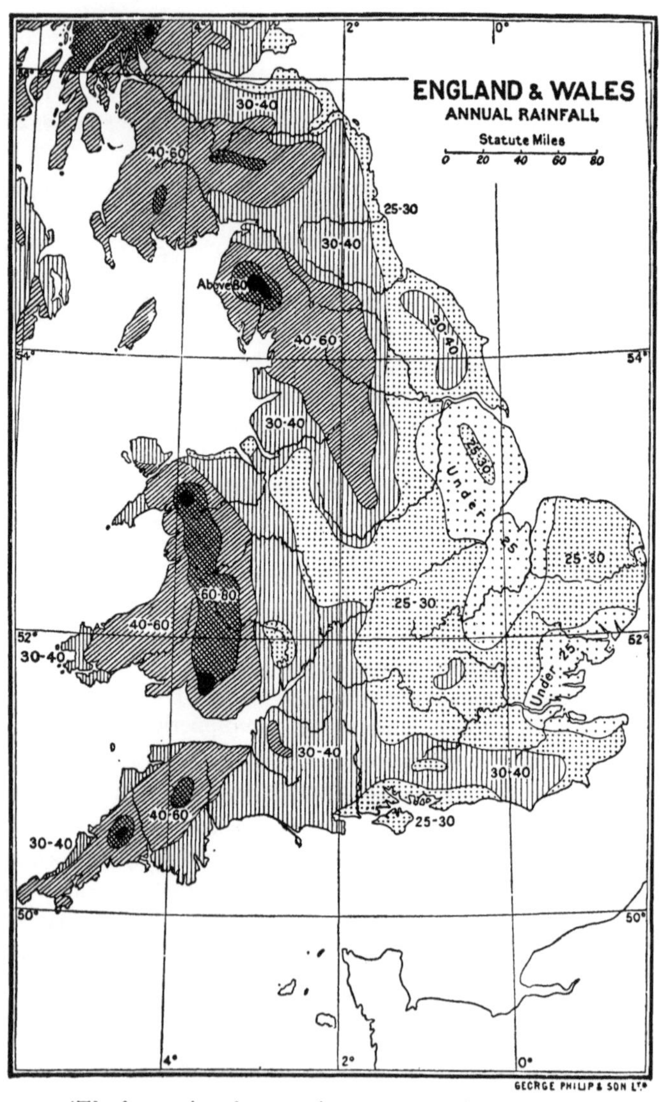

(The figures give the approximate annual rainfall in inches)

rainfall of England on the opposite page, where it will at once be noticed that the heaviest fall is in the west, and that it decreases with remarkable regularity until the least fall is reached on our eastern shores. Thus in 1908, the maximum rainfall for the year occurred at Llyn llydaw in the Snowdon district, where 237 inches of rain fell; and the lowest was at Bourne in Lincolnshire, with a record of about 15 inches. These western highlands, therefore, may not inaptly be compared to an umbrella, sheltering the country farther eastward from the rain.

The above causes, then, are those mainly concerned in influencing the weather, but there are other and more local factors which often affect greatly the climate of a place, such, for example, as configuration, position, and soil. The shelter of a range of hills, a southern aspect, a sandy soil, will thus produce conditions which may differ greatly from those of a place—perhaps at no great distance—situated on a wind-swept northern slope with a cold clay soil.

If a range of hills lies across the onward path of moisture-laden winds, the rainfall is largely increased on the side facing the winds, and reduced over the country on the other side of the range. This is evident in Oxfordshire. The south-westerly winds bring most rain, on account of their long sweep over the Atlantic; but the rain clouds discharge their moisture over the western hills, and therefore when the winds reach the plains of Oxfordshire they are drier, as the air descends to lower levels. There is, for this reason, more rain on the Edge Hills and in the Chiltern district than in other parts of the shire, and the rainfall in the central district and in the north is fairly small.

Inhabitants of Oxford always speak disparagingly of its climate. A damp dull day with showers and fog is always regarded as typical Oxford weather, but unjustly. The rainfall is not remarkably heavy, though the proximity of the river Thames tends to keep the atmosphere moist. The climate of the county is for the most part salubrious and dry, but in winter colder than the other southern districts of England, especially in the bleak and exposed regions of the Chilterns, though in summer it is warmer.

The Radcliffe Observatory at Oxford is the only organised meteorological station of the first class in the county, and though there are now several other stations, there is a dearth of reliable weather statistics elsewhere in Oxfordshire. The Observatory, erected in 1772, takes its name from the munificent Dr Radcliffe, who built also the Radcliffe Library and the Radcliffe Infirmary, and was a benefactor to University College. It has an octagonal tower designed from the Temple of the Winds at Athens and surmounted by a large globe supported by figures of Atlas and Hercules. The Observatory is fitted with the best modern astronomical instruments, and the height of the barometer and thermometer, the direction of the winds, and the state of the weather are registered continuously by an ingenious apparatus of photography.

The average rainfall at Oxford, the mean of 94 years, is 26·013 inches, and it may be concluded from reasons stated above and from actual observations, that the rainfall on the Chilterns is about two inches higher, and that there is also an increase in the hilly regions of the northern part of the county.

The average temperature, the mean of 81 years, at Oxford is 48°·88, which is rather above the average (48°) for the whole of England. The county, however, is not conspicuous for its sunshine. Our southern and eastern coast-towns generally show the highest average in this

Radcliffe Observatory, Oxford

respect, 1800 hours out of the possible 4435 hours the sun is above the horizon being often recorded, and sometimes nearly 2000. But its inland position and liability to fogs give Oxfordshire a less favourable record, and the average hours of bright sunshine in Oxford, from a mean of 28 years, is only 1465 hours. The sun thus shines at Oxford about one-third of the possible maximum number of hours, and though we have no reliable statistics from places further away from river fogs, we may conclude that it is a little brighter there.

9. People—Race, Dialect, Settlements, Population.

Oxfordshire at the time of the Roman conquest of Britain was inhabited, as we have said, by the Dobuni, a warlike Celtic tribe who had for their neighbours other powerful tribes, whose names have already been recorded. Amongst them were the Catuvelauni on the west (familiar to the readers of Tennyson as the Catyeuchlanians) with whom the Dobuni were continually at war, and who, at one time, held all the district, making the Edge Hills their western boundary. The Dobuni submitted themselves to Aulus Plautius at Cirencester. Oxfordshire has few Roman towns, though it could show many fine villas, the residences of powerful Romans. During the four centuries of Roman rule, the history of the district is a blank. Doubtless many of the Britons lingered on, slaves and servants of the Roman lord, or as wild outlaws in the great woods, ever ready to raid or attack a lonely farm. But the Romans disappeared and the Saxons came, and thoroughly colonised the district.

The Celts however left traces of themselves behind them in the names of the rivers. The words Thames or Temese, meaning broad water, Thame, Isis, Evenlode, or Avonlode, Ray, Cherwell, are all Celtic, and also

Dorchester, the city of the Dur-otriges, or dwellers by the water.

The Saxons thoroughly colonised the district and by far the greater number of place-names declare their origin. All words ending in *ton*, or *ham*, or *field*, or *ford*, are Saxon, and were first formed as settlements of Saxon families and clearings in the forests. Oxfordshire is a Saxon county, and the people retain many of the characteristics of their ancestors. We are, however, not without some traces of the Danish conquerors. *Thorpe*, or *throp*, meaning a village, betrays their existence, and the words Heythrop, Dunthrop, Thrup, near Woodstock, and Cokethorpe, show that they had some settlements or villages in the county. Oxfordshire folk are chiefly descendants of the Anglo-Saxons.

The great and powerful Norman families who came over with the Conqueror or subsequently rose to power have given their names to some villages and estates. Thus we find the Harcourts at Stanton Harcourt, the St Johns at Stanton St John, the Baldwins at Brightwell Baldwin, the Peppards or Pypards at Rotherfield Peppard, the Greys at Rotherfield Greys. Sometimes the Normans gave a name to a place, which was afterwards corrupted and Anglicised. Thus the hill outside Oxford they called *chateau vert*, but it is difficult to recognise in that name the origin of the Shotover of to-day.

There does not seem to have been any extensive immigration of foreigners into Oxfordshire. At the university there were many students of other nations, and foreign teachers and doctors often taught in the schools of the university. But all these were migratory, and had little effect on the population of the county. There is, however, evidence that Flemish weavers came to Witney to improve the art of the clothiers, and the family of Brabant is especially mentioned in the records of that town.

The county, being mostly agricultural, has suffered

An Oxfordshire Farm

from the depression which has tried that industry, and the population has decreased. Young men in these days find the country dull and fly to the towns for employment and amusement. It does not always happen that they are better off in towns, as the rents and cost of living are dearer, and they have no gardens. When the last census of 1911 was taken, the population of the administrative county was 199,277 persons. Ten years earlier it was 186,460. At the beginning of the nineteenth century it was 111,977; so there has been an increase during the last 100 years. The great decrease of the population is in the country villages. The suburbs of Oxford has increased enormously along the Woodstock and Banbury roads, and during the last century has added 37,057 to its inhabitants. Caversham has become a suburban portion of the Berkshire town Reading; its inhabitants increased from 3583 in 1881 to 6580 in 1901, and at the last census the numbers have risen to 9858. Along the Thames' banks many houses have been built, but the old-fashioned market towns, Chipping Norton, Bicester, Thame, and Witney, have remained almost stationary in population.

The census shows that in Oxfordshire the females exceeded the males by 3172.

10. Agriculture — Main Cultivations, Stock, Woodlands.

As agriculture is the most important industry of the county, it will be interesting to study some particulars concerning it. A large proportion of the land is under cultivation. Great Britain has 56,200,000 acres, and of these 32,243,447 acres are being cultivated. The administrative county of Oxford contains 475,968 acres, including water, and of these 411,874 acres are under cultivation, about eight-ninths of the total area of the county. But of this more than half, or 208,524 acres, are permanent pasture.

The soil of Oxfordshire differs considerably during its length of 50 miles. We have 80,000 acres of red land in the regions of Banbury, Hook Norton, Adderbury, and Wigginton, which is good farming land, neither too light nor too strong; 166,000 acres of stonebrash round Blenheim, Chesterton, Witney, Burford, and Charlbury, which is excellent for sheep and barley; 66,000 acres of flinty-covered ground about the Chilterns, which is not so good, but the flat arable fields round Reading are profitable, except when they are burnt up in hot dry summers. Besides all these, there are many acres of varying soil, which are used for grazing land and milking pastures.

The corn crops of Oxfordshire consist of wheat, barley, oats, beans, and peas. Rye is not much grown in the county. Wheat crops occupy 38,638 acres, about half the area used for this crop 40 years ago. This is owing to the great decrease in the price of wheat, and to the large amount imported, for some 28,000,000 quarters of wheat, in addition to much flour, are sent to this country every year from abroad. Recently the price of corn has been steadily advancing; hence there has been a slight increase in its cultivation, and it is to be hoped that this will bring some return of prosperity to the farmers. In 1907 wheat crops covered 32,968 acres; in 1908 the acreage increased to 33,703, in 1909 to 38,638.

Barley occupies 34,966 acres, whereas 40 years ago it required 52,069 acres for its cultivation. Oats, on the other hand, have greatly increased. In 1867 the acreage under this cereal was 22,862; in 1908 it had gone up to 34,485, but in 1909 it had dwindled to 32,210.

Vetches, turnips, and swedes are largely grown. As we shall see presently, Oxfordshire farmers breed many sheep and cattle, which depend for their subsistence on these crops of turnips and swedes, as well as on corn, cake, and hay. The sheep are penned in hurdles on the fields

Oxford Down Ram

and consume the roots on the ground, thus tending greatly to manure and enrich the land. Mangolds have risen in favour recently for the feeding of sheep and are much cultivated.

The soil of Oxfordshire is suitable for potatoes, the acreage of which has doubled during the last 40 years; and lucerne, sainfoin, red clover, as well as white, and trefoil are much sown.

The Oxfordshire live-stock requires little description except the sheep, which are an important feature. Farm horses have recently improved in quality, but everywhere the increased use of machinery for farming purposes tends to diminish their number, and motors and steam-tractors will, perhaps, render the breed of horses extinct. Short-horn cows are usually found on most of the farms, and sandy-coloured pigs, a cross between the Tamworth and the Berkshire breeds. The Oxfordshire sheep are famous, and are known as the "Oxford Down" or "Down Cotswold," a cross between a Cotswold and a Hampshire. They produce splendid fleeces of wool and good mutton, and there is not a better breed in England than the Oxfordshire sheep, which an old agricultural writer calls "the glory of the county." A recent observer has said of them:—"The sheep so gross, so superb, so immense, have been commemorated by many artists of an earlier day. You can hardly enter a farm-house or an inn without seeing these unwieldy objects limned for your admiration, and you wonder how such small legs could support so weighty a fleece. But it is these same sheep that gave the Cotswolds their old prosperity, and made Northleach and Burford, and such towns rich in their wool-staplers, who belonged to the staple of Calais." The same writer speaks of the Cotswold dog as "a great woolly creature with very little of a tail, and his gambols remind one of Macaulay's hippopotamus." But he is a faithful animal and most intelligent in the driving and herding of sheep.

In spite of the destruction of woods and forests, there

are still 18,528 acres in the county covered with timber, and 763 acres have been planted during the last ten years. People have begun at last to pay some attention to growing trees; we used to content ourselves with cutting down the trees which our ancestors planted, and never planting others to take their places. But during the last ten years nearly half a million acres have been planted throughout

Worcester College Gardens

England, though the area of Oxfordshire woodland during that period has decreased, and no less than 6000 acres of woods and plantations have been shorn of their trees since 1905. There are some grand trees and magnificent parks in the county, and many of the colleges have beautiful gardens. In Holton Park there is an oak measuring 28 feet in circumference, just above the ground, an elm

28 feet 4 inches, and a beech 19 feet. Dr Plot mentions an oak near Nuneham Courtenay that shades 460 square yards, beneath which 2420 men could shelter themselves from rain or sun. At Magdalen College, Oxford, there was an oak that shaded 768 square yards, and one at Rycote that shaded 972 square yards. On Kidlington Green was a giant hollow oak that was used for imprisoning vagabonds for a night before they were removed to Oxford gaol, and eight or ten could be conveniently housed, the tree being 25 feet round.

11. Industries and Manufactures.

The county has several important industries and manufactures, some of which depend largely for their support on the needs of Oxford University. It has few natural products to aid it in its industrial enterprise. As already said, the Thames river was a valuable asset in the days before railways were invented. It opened the London market for the goods of Oxford and Henley, from which place malt was sent to the metropolis in very early times. The pure waters of the Cherwell induced the leather-dressers to establish their industry at Oxford in the seventeenth century, and those of the Windrush doubtless partly accounted for the excellence of the Witney blankets. The rich fleeces of the Cotswold sheep made this wool-trade prosperous, and cloth-weaving was extensively carried on in most of the towns and villages of the shire until steam and coal turned the fortunes of the textile industry elsewhere. An important natural product is the building-stone which abounds in many parts. Clay for making bricks is plentiful, and the iron ore which was discovered in the northern part of the county 50 years ago has been extensively worked, and in one year as much as 36,808 tons, worth £7721, have been dug up. The ore is sent to the furnaces of South Staffordshire and North Wales. These are the only industries that arise out of the natural products of the county.

The presence of the University at Oxford has called into being the University Press, which is the largest concern in the city, employing several hundreds of work-people. It dates back to the dawn of printing in England, its earliest book, a treatise on the Apostles' Creed, having been printed in 1468, or as most scholars think in 1478. The history of the Press has not been continuous. There have been breaks and periods when there was no printing press in Oxford. In 1517 printing was revived, continued for a few years, and then ceased. In 1585 the Oxford Press again started, and has continued ever since to produce the copies of the Holy Scriptures and works of learning for which it is famous. Countless numbers of books have issued from this Press. When the Revised Version of the New Testament was issued on May 17th, 1881, a million copies were sold on the first day. Since its first commencement in 1478 to the beginning of the present century it has printed 19,475 works, and 9800 of these were issued during the latter half of the nineteenth century. The printing was for over a century

Oxford University Press

carried on in the Clarendon Building in Broad Street, which was erected out of the profits of Lord Clarendon's *History of the Rebellion*.

Next to the printing of books comes their clothing and binding, and this industry was carried on in Oxford long before types and printing presses were invented. One Laurencius was a bookbinder at Oxford at the end

Room in the Bodleian Library

of the twelfth century, and since that time the names of many binders appear in the city records. When Thomas Bodley founded the famous Bodleian Library, in the first year of James I, there was an enormous increase in the trade, and it has since flourished in the hands of many distinguished and accomplished binders. Another branch of book production is the trade of parchment and paper making. One Reginald, a parchment maker, lived at Oxford at the close of the twelfth century, when Laurencius was binding books. These tradesmen, with others of their calling, lived in Cat Street, at a time when the members of a particular calling always were required to reside in a certain row or street. Parchment-making

Wolvercote Paper Mills

flourished until the end of the sixteenth century, and then the making of paper began. Several mills for making paper have existed in the county and still carry on their trade. The most famous is that at Wolvercote, started in 1666, which continues to produce some of the best paper in England, and is now owned by the University Press. Another mill exists at Eynsham, where leather boards are now made. Sandford mill, once a corn mill, has a long history, and produced coloured papers. The Weirs mill near Oxford, and several near Henley, have also contributed to this industry.

Cloth-weaving existed at Witney as early as 969, and at Oxford in 1130. At least these are the earliest records. Burford, Chipping Norton, and Banbury, as well as Oxford, were flourishing centres of the industry in the sixteenth century. William Stumpe, who converted Malmesbury Abbey into a cloth factory, wished to do the same with Osney Abbey, but owing to the conditions required by the corporation, his scheme was abandoned. In the eighteenth century the trade of the weavers flourished, and in many of the villages hand-loom weaving was carried on.

Banbury was chiefly famous for making worsted plush, and still carries on that trade. Web and girth-making are also carried on there.

Horse-cloths are still made at Chipping Norton, but modern methods with the use of coal and steam have silenced the looms of Oxfordshire, and transferred the trade to the manufacturing districts of Yorkshire and Lancashire. The weaving of silk stockings, and silk-"throwing" or winding, was carried on in the county at Oxford, Banbury, and Henley, but this too has vanished. On the borders of Buckinghamshire lace is made in many cottages, the industry having improved from the revival which has taken place in recent years.

The leather-trade once flourished in many parts of the county, and Burford was famous for its saddles.

On two occasions it had the honour of presenting specimens of the skill of its saddlers to royalty—on the visit of Charles II to the town in 1681, and that of

Cake Shop, Banbury

William III in 1695. Gloves are made at Woodstock and the industry is very ancient, having certainly existed in Saxon times.

Marsh Mills, on the Thames

(On the left-hand is the Berkshire bank)

Malting and brewing have flourished in several Oxfordshire towns and the industry is still widely carried on, as is also the manufacture of tiles and bricks.

The requirements of modern agriculture have created the extensive industry of making agricultural machines for which Banbury is famous, and there are large works at Cowley. Bell-founding still exists at Burford, where in the church there is a Bell-founders' Aisle, and where, in the reign of Charles I, Henry Neale was a celebrated founder. Oxford and Woodstock also contributed their share to the bells of the county.

The presence of the great river naturally has brought into existence the industry of boat-building. Old barges were made in former days, and now the names of Salter and Clasper are famous all the world over for their splendid racing craft, as well as for their motor, electric, and steam launches. Boats are also built at Goring, Shiplake, and Henley.

Chairs are made largely at Stokenchurch, which was formerly in the county, and also at Chinnor, Watlington, and Caversham. Banbury is still famous for its cakes, which can claim a very respectable antiquity, as they were in favour in the time of Queen Elizabeth; and formerly it was noted for its cheese, the fame of which has now vanished.

12. A Special Industry — Witney Blankets.

There is one manufacture for which Oxfordshire is especially noted, and that is the making of blankets. All the world knows of Witney blankets and some account of the industry is necessary here. It is certainly ancient. Even before the Norman Conquest the wool trade was carried on at Witney. The river Windrush afforded excellent water for cleansing the wool, the Cotswold sheep supplied rich fleeces in abundance, and Witney possessed every advantage for the trade. But who first invented blankets? A story is told of one Thomas Blanket of Bristol, who is said in 1320 to have discovered the process of making these most useful coverings, and gave his name to the material he invented. That story may be true, and it may not. It was probably invented to account for the use of the word, which is really derived from the French *blanchet* (a white thing) and was in use in England in the time of Henry II. We do not hear much of Witney blankets before the reign of James I, though the woollen industry had prospered there for a long time. But after that period the trade in blanket-making grew and prospered greatly, Witney blankets finding their way to all parts of England and even to the natives of Africa. The Wenman, Early, and Brookes families were the chief manufacturers in the sixteenth and seventeenth centuries. In 1710 the Witney merchants gained a royal charter, granted to the Company

Witney Blanket Factory

of Blanket-weavers. They built themselves a Hall, called Blanket Hall, and quarrelled much among themselves over by-laws and agreements. At the beginning of the nineteenth century the trade was very prosperous. Three thousand hands were employed; Kersey-pieces, bear-skins, and blankets were exported to America, Spain, and Portugal. Machinery was introduced and added greatly to the output. Almost 93,000 blankets were made every year.

The fame of the Witney blankets continues and the trade is very prosperous. Eight hundred hands are employed and 250 looms. Wools are brought to the factories not only from England, but also from Australia, New Zealand, India, and other countries. This wool has to pass through many processes before it becomes a blanket. The wools are first blended together, and then passed through various machines called willeys, teazers, scribblers, and corders, which break up the wool and reduce it to a condition such that it can be wound on bobbins. Then the threads are stretched and twisted on the spinning mule, and are ready for weaving. When woven the material goes to the fulling mill and is beaten by heavy hammers. After that it is washed and dried and bleached and stretched, and last of all the fibres of the wool are drawn out from the surface of the cloth by a machine. And thus we get our Witney blankets.

13. Quarries and Minerals.

With the exception of the iron ore, which is found in the northern district, an account of which has already been given, there are no important minerals in Oxfordshire. But its stone quarries are numerous and have had a great history. All the beautiful colleges of Oxford and the noble churches that abound in the county, are built of local stone taken from these quarries. It is not possible that the founders of Oxford University selected its site on the Thames because they were aware of the masses of stone which the neighbouring hills afforded; nor did they anticipate the erection of such beautiful and extensive college buildings as subsequently arose. But they were certainly fortunate in their choice. The county has numerous quarries of freestone, limestone is abundant, and slate can be obtained in several places.

In Anglo-Saxon times some of the stone was quarried. The tower of St Michael's Church, Oxford, is constructed of stone from the quarry at Chilswell. So also are the remains of Norman work in the city. Teynton stone was also used in early times. Headington and Taynton quarries supplied all the stone for the fifteenth century colleges and for Thame Church, and the Burford and Holton quarries supplied their share for Oxford needs. Handborough and Woodstock opened quarries in the seventeenth century and Stonesfield supplied slates. The stone for rebuilding St Paul's Cathedral is said to have been brought from Burford. This latter is harder and whiter than the stone of Headington, which is liable to decay and to the influence of the weather. Other quarries existed at Bladon, Little Milton, Barford, and Hornton. Blenheim Palace was built for the most part of Taynton stone. Some coarse marble is found near Banbury and in Wychwood Forest. It has been found necessary to reface parts of the walls of the colleges owing to the decay of the Headington stone, and the Milton quarries have supplied most of the material for this work. There are still about 40 quarries in the county, but only about 130 quarrymen. People now prefer to make their houses of cheap bricks, and do not care to build so surely and so well as our ancestors. Hence the quarries are neglected and the industry is decayed.

Brickmaking flourishes, and there is an abundance of clay in the county. Kilns and brickworks exist near Oxford, especially at Wolvercote and Summertown, also near Bicester, at Finmere, Goring, Long Handborough, Caversham, Nettlebed, Wheatley, Culham, Banbury and elsewhere. The clay at Shotover was used for making tobacco-pipes for the King's soldiers when they were quartered at Oxford during the Civil War.

There are plenty of gravel-pits, especially near the course of the Thames and on the southern slopes of the Chiltern Hills. Flints picked from the surface of the fields on the Chilterns furnish excellent material for road-making when the traffic is not too heavy.

14. History of the County.

On the coming of the Romans, Caesar could not penetrate the forests of Oxfordshire, and it was left to Aulus Plautius nearly a century later to receive the submission of the Dobuni. After the departure of the Roman legions the Anglo-Saxons came to Britain, and about the middle of the sixth century they found their way to Oxfordshire. After taking Silchester they fought with the Britons in 556 at Beranbyrig, probably Barbury, with only partial success. Cuthwulf and his West Saxons in 571 took Bensington, Aylesbury, and Eynsham, and eventually overran the country. A long period of anarchy ensued. The West Saxons advancing from the south were confronted with the Mercians coming from the north, and the records of six centuries tell of little but constant fighting between this opposing power and their constant enemy, the Danes. Wulfhere, King of Mercia, invaded Berkshire and took possession of its northern part. In 752 Cuthred of Wessex crossed the Thames, fought Ethelbald of Mercia at Burford, and conquered the country. "Battle Edge," near Burford, marks the site of the battle, which was commemorated as late as the eighteenth century by a feast and a procession, when the figures of a dragon and a giant were carried through the streets.

Offa of Mercia reconquered his lost possessions by winning the battle of Bensington in 777 A.D. But Egbert of Wessex established his sway over Oxfordshire in 827.

Christianity came to the West Saxons by means of St Birinus, who converted Cynegils, King of Wessex, preaching to him at Churn Knob on the Berkshire downs. Oswald, the Christian King of Northumbria, was with him, and Cynegils was baptised at Dorchester,

Dorchester Church

which was given to St Birinus as the seat of his bishopric. This little Oxfordshire town became the episcopal seat of a diocese extending from the Thames to the Humber. The name of the first bishop still survives in Berin's Hill on the Chilterns.

Oxfordshire suffered severely from the Danes. In

914 they plundered as far as Hook Norton, and destroyed many towns and villages. Edward the Elder checked their ravages. Kirtlington was the scene of a great council in 977. When the Danes renewed their ravages Ethelred the Unready tried to buy them off, and then on St Brice's Day, November 13, 1002, ordered their massacre at Oxford. In revenge they ravaged the country and burnt Oxford. The Saxon kings had three palaces in the shire, Woodstock, Headington, and Islip. Gemots were held in Oxford in 1018 and 1036, and it was there Harold died. Edward the Confessor was born at Islip, which he gave to the monks of Westminster.

After the Norman Conquest, Robert D'Oilly was the most important person in the shire; he married the heiress of Wigod, lord of Wallingford, and held 50 manors in the county. He was ordered to build a castle at Oxford. Castles were built during Stephen's reign at 14 places, and the extensive forests invited the Norman kings to hunt the deer. Henry I had a hunting lodge at Woodstock, and built the palace of Beaumont just outside Oxford. The civil war of Stephen's reign raged in the county. The castles of Oxford, Woodstock, and Bampton were held by the Empress Maud, who was besieged and hard pressed at Oxford, and escaped by night along the frozen Thames dressed in white, finding safety in the fortress of Wallingford. A council at Oxford in 1153 ended the war.

Henry II destroyed several of the castles erected in the reign of his predecessor. Woodstock was famous for the story of Fair Rosamond, who was buried at Godstow,

Minster Lovell

and at Woodstock first arose the storm that raged between Archbishop Becket and the King.

Richard I and John were both Oxfordshire men, the former having been born at the palace of Beaumont, the latter at Woodstock. Many parliaments and councils were held at Oxford, and there the "Provisions of Oxford" were drawn up in 1258, which rank with Magna Charta as the safeguard of English liberty. A notable figure in the history of the county in the thirteenth century was Richard, King of the Romans, who had a palace at Beckley. Piers Gaveston was imprisoned at Deddington Castle just before his death at the hands of the "Black Dog," Earl of Warwick.

Fighting took place at Radcot Bridge in 1387, when Robert de Vere, Earl of Oxford, a favourite of Richard II, fought against the forces of the Earl of Gloucester and Derby. At Oxford in 1400 a conspiracy was made to murder King Henry IV at a tournament and to proclaim a certain priest of Magdalen College dressed in royal robes as Richard II returned to life. The conspiracy failed and many noble heads fell.

The Wars of the Roses affected the life of the county. Romance states that in Wychwood Forest Edward IV first saw and loved Elizabeth Woodville. The battle of Danesmoor, near Banbury, was fought in this reign between an army of insurgents from the north and the royal forces led by the Earl of Pembroke. The insurgents won, and beheaded the Earl and other leaders at Banbury. The fall of Richard III sealed the fate of several Oxfordshire families, and amongst them the Lovells of Minster Lovell.

The Reredos, All Souls' Chapel

Many changes took place during the period of the Reformation. The ecclesiastical architecture of Oxford suffered at the hands of the image-breakers, and it was not until the last century that the plaster was removed from the reredos of All Souls' and the figures restored to their present condition. With the spoils of the monasteries

Balliol College and the Martyrs' Memorial

Henry VIII founded six bishoprics, of which Oxford was one. The Oxfordshire rustics revolted on account of the introduction of the new Prayer Book, and an army of 1500 men was sent against them. Ridley, Latimer, and Cranmer were burnt at Oxford in Mary's reign. Princess Elizabeth during the reign of Mary was a prisoner at Rycote and Woodstock. The Queen's favourite, the

Earl of Leicester, died at Cornbury. There were several families of Recusants[1] in the county, and amongst them the Stonors of Stonor Park, who sheltered Campion the Jesuit and allowed him to set up a secret printing press.

Oxfordshire played an important part in the Civil War, for Oxford was the royalist headquarters. Battles were fought at Cropredy and Chalgrove, and at Edgehill, just beyond the borders of the county. The castle of Banbury was a great centre of fighting, and the country around was pillaged by both belligerents. A little room at Broughton Castle, the seat of Lord Saye and Sele—"Old Subtlety"—is pointed out as the birthplace of the rebellion. The whole county was a battle-field, and skirmishes took place everywhere. The havoc wrought by the war was terrible. Towns were pillaged, old houses destroyed, Banbury Castle pulled down, and desolation reigned.

The rebellion of the Levellers troubled the Commonwealth, and Cromwell took stern measures at Burford to crush the insurrection, three of the leaders being shot in the churchyard. The marks of the bullets on the churchyard wall and the name "Anthonye Sedley 1649, prisner," cut on the lead font in the church wherein the Levellers were confined are still to be seen.

When the Plague broke out in London, Charles II repaired to Oxford. Parliament met there in 1681.

[1] A Recusant was one who adhered to the Roman Catholic religion and refused to accept the Acts of Uniformity passed under Elizabeth and succeeding sovereigns.

The efforts of James II to Romanise the University met with the stubborn defence of the Fellows of Magdalen. The city sympathised with the fallen house of Stuart, and Wesley said it was "paved with the skulls of Jacobites." The squires of the county meditated joining the rising of

Blenheim Palace

1745, and the Pretender is said to have visited Lord Cornbury and to have been shaved by a barber of Charlbury.

Blenheim Palace was given by a grateful nation to the hero of many fights, the great Duke of Marlborough. With the dying flickers of the flame of Jacobitism and the honour bestowed upon a famous general, the connection of Oxfordshire with the annals of England may be said to have ceased.

15. Antiquities—Prehistoric, Roman, Saxon.

The prehistoric period of Britain may be divided into four periods: (1) the Palaeolithic or Old Stone Age; (2) the Neolithic or New Stone Age, between which periods in England a great gap of time existed; (3) the Bronze Age; (4) the Iron Age. The people who dwelt in these periods have all left traces behind them. In the old history books the story of Britain began with the landing of Julius Caesar, B.C. 55; but the discoveries of the last half century, the finding and classification of the weapons and tools of flint or bronze, the exploration of barrows, lake-dwellings, earthworks, and cromlechs, have pushed back our historical horizon and enabled us to know the manner of life of the tribes and races who dwelt in Britain centuries before the Roman invasion.

The traces of Palaeolithic man, who lived mainly if not entirely in the more southern part of our land, are very numerous, and he evidently exercised great skill in bringing his implements to a symmetrical shape by chipping. Stone, wood, and bone were his only materials. He lived here during the period when this country was united with the continent, and when the huge mammoth roamed in the wild forests, and powerful and fierce animals struggled for existence in the hills and vales of a cold and inclement country. His weapons and tools were of the rudest description, and made of chipped flint. Eighty or ninety feet above the present level of the Thames at Caversham, in the higher gravels, are these relics found, and they are so abundant that the race must have been fairly numerous. The shape of the weapons is usually oval, and often pointed into a rude resemblance of the shape of a spear-head. Some flint-flakes are of the knife-like character; others resemble awls or borers with sharp points, evidently for making holes in skins for the purpose of constructing a garment. Hammer-stones for crushing bones, tools with well-wrought flat edges, scrapers, and other implements, were the stock-in-trade of the earliest inhabitants of our country, and are distinguishable from those used by Neolithic man by their larger and rougher work. An interesting find was made at Caversham—the bones of a mammoth surrounded by a large number of flint weapons, showing that their owners had attacked and killed the monster with these primitive weapons. During this time, the elk and reindeer, the gigantic Irish deer, bison, elephant, rhinoceros, hippopotamus, lion, bear and other creatures roamed this country-side.

After a lengthy period of geological change, Neolithic man appeared, probably peopling the country from the south and east, a much more civilised person than his predecessor and presenting a higher type of humanity. He had a peculiarly shaped head, the back part of the skull being much prolonged, and from this feature he is called *dolichocephalic*. He discovered that flints that were dug up were much more easily fashioned than those which lay on the surface of the ground. He polished his weapons and fashioned finely wrought arrow-heads and

1. Palaeolithic and 2. Neolithic Implements

javelin-points. He made pit-huts to dwell in, cultivated the ground, and had domestic animals. The long barrows or mounds, the length of which is greater than the breadth, contain his remains. We find such long barrows at Lyneham and in Slatepits Copse in Wychwood Forest.

Another wave of invaders swept over the island and conquered the Neolithic race. These were the Celtic people, taller and stronger than their predecessors, and distinguished by their fair hair and rounded skulls. From the shape of their heads they are called *brachycephalic*, or short-headed, and are believed to have belonged to the original Aryan race, whose birthplace was Southern Asia.

The Rollright Stone Circle

Their weapons were made of bronze, although they used polished stone implements also. As they became more civilised they discovered the use of iron, of which they fashioned axe-heads. Their remains lie in the round barrows, of which there are no less than 37 in the county.

The Rollright stone circle, which stands just on the county boundary about half a mile north-north-east of

Little Rollright, is the most important megalithic monument in Oxfordshire, and has been described as the second wonder of the realm—second, that is, to Stonehenge. It has been used as a quarry and many stones have disappeared, but we can still easily trace the circle. Near it stands the "King's Stone" and, not far off, others called the Whispering Knights. A legend states that a chieftain, relying on an old prophecy which stated that if he could once see Long Compton he would be King of England, marched his army in that direction. While he was repeating the words

> "If Long Compton I can see
> King of England I shall be,"

Mother Shipton appeared and pronounced the spell,

> "Move no more: stand fast stone;
> King of England thou shalt be none."

The whole company were then turned into stone. The solitary stone is the King, the circle his army and the Whispering Knights are some conspirators who were plotting against the King. This is the story. But, of course, the real object of circles such as this and Stonehenge was either monumental, or ritual. Dolmens or table-stones, which were sometimes covered with earth, were undoubtedly sepulchral. There is a fine example of these, called the Hoar Stone, at Enstone. Sometimes single stones were erected, such as the Hawk Stone and Thor Stone, also near Enstone. Frethelestone has been broken up, but we have the Devil's Quoits, near Stanton Harcourt, once a circle of stones. Popular tradition states that the Devil once played quoits with a beggar for his soul and won by flinging these great stones. All these stone monuments were probably raised by the Bronze Age people, and were for the object of marking

Rollright Circle: the King's Stone

the graves of illustrious chiefs. Avebury and Rollright, made of undressed stones, are more ancient than Stonehenge, which has immense trilithons, or stones arranged in the shape of a doorway. It is not improbable that the Rollright circle was formed as much as twenty centuries before Christ. The relics of prehistoric man have been found in all parts of the county, and his burial mounds abound all over the portion west of the Cherwell.

Great attention has been paid in recent years to the study of earthworks, camps, and fortifications. These have now been grouped for description on a regular and scientific basis. The district of the county west of the Cherwell contains most of the early earthworks, which cluster most in the north-west of this part. Madmarston, Tadmarton, Idbury, Lyneham, and Chastleton belong to the Cotswold series. There are several lines of entrenchments in England, usually called dykes, and three of them exist in Oxfordshire. There is Grim's Ditch, or Grimes Dyke, on the north of Akeman Street, another Grim's Ditch in the south of the county between Mongewell and Henley, 11 miles in length, with "Madder's Bank," near it and parallel to it, some 14 miles in length. There is yet another entrenchment—called by various names, Aveditch or Avesditch, Ashbank, and Wattlebank—extending from where the Akeman Street crosses the Cherwell to the northern boundary of the county at Souldern. Although the point is still undecided it is conjectured that these were of Roman construction, the northern ones made to defend the Akeman Street, the southern one to protect the Icknield Way, this latter being an old British track-way, afterwards used by the Romans. Finally, mention of the British village which exists at Standlake must not be omitted. This was discovered in 1857, when thirteen hut-circles were explored, and many of the objects obtained placed in the

Ashmolean Museum at Oxford. No remains of the hut-circles are now visible.

The county was not so thoroughly colonised by the Romans as some other parts of England. Perhaps the dense forests prevented them. Roman stations existed at Alchester and Dorchester, but the county was dotted over with Roman villas, some of them the finest in England. Stonesfield, Northleigh, Beckley, Wheatley, Fringford, Middleton Stoney, and Woodperry possessed good examples of the residences of noble Romans of which little now remains to be seen.

Saxon remains are by no means rare. In the barrow at Lyneham there were the remains of a Saxon burial, with javelins, knife, and the umbo or boss of a Saxon shield.

16. Architecture—(*a*) Ecclesiastical.

A preliminary word on the various styles of English architecture is necessary before we consider the churches and other important buildings of our county.

Pre-Norman or, as it is usually, though with no great certainty termed, Saxon building in England, was the work of early craftsmen with an imperfect knowledge of stone construction, who commonly used rough rubble walls, no buttresses, small semicircular or triangular arches, and square towers with what is termed "long-and-short work" at the quoins or corners. It survives almost solely in portions of small churches.

The Norman Conquest started a widespread building

of massive churches and castles in the continental style called Romanesque, which in England has got the name of "Norman." They had walls of great thickness, semi-circular vaults, round-headed doors and windows, and massive square towers.

From 1150 to 1200 the building became lighter, the arches pointed, and there was perfected the science of vaulting, by which the weight is brought upon piers and buttresses. This method of building, the "Gothic," originated from the endeavour to cover the widest and loftiest areas with the greatest economy of stone. The first English Gothic, called "Early English," from about 1180 to 1250, is characterised by slender piers (commonly of marble), lofty pointed vaults, and long, narrow, lancet-headed windows. After 1250 the windows became broader, divided up, and ornamented by patterns of tracery, while in the vault the ribs were multiplied. The greatest elegance of English Gothic was reached from 1260 to 1290, at which date English sculpture was at its highest, and art in painting, coloured glass making, and general craftsmanship at its zenith.

After 1300 the structure of stone buildings began to be overlaid with ornament, the window tracery and vault ribs were of intricate patterns, the pinnacles and spires loaded with crocket and ornament. This later style is known as "Decorated," and came to an end with the Black Death, which stopped all building for a time.

With the changed conditions of life the type of building changed. With curious uniformity and quickness the style called "Perpendicular"—which is unknown abroad—developed after 1360 in all parts of England and lasted with scarcely any change up to 1520. As its name implies, it is characterised by the perpendicular arrangement of the tracery and panels on walls and in windows, and it is also distinguished by the flattened arches and the square arrangement of the mouldings over them, by the elaborate vault-traceries (especially fan-vaulting), and by the use of flat roofs and towers without spires.

The medieval styles in England ended with the dissolution of the monasteries (1530–1540), for the Reformation checked the building of churches. There succeeded the building of manor houses, in which the style called "Tudor" arose—distinguished by flat-headed windows, level ceilings, and panelled rooms. The ornaments of classic style were introduced under the influences of Renaissance sculpture and distinguish the "Jacobean" style, so called after James I. A fine example of this is seen in the second quadrangle of St John's College. About this time the professional architect arose. Hitherto, building had been entirely in the hands of the builder and the craftsman.

Although the churches of Oxfordshire are not equal in size and beauty to those of Northamptonshire or the Fenland, no county can claim to possess a series of churches of greater general interest and special architectural excellence than that which is here described. With very few exceptions the churches are of mixed styles. The first Norman lord or Saxon thane began to build a small church on his estate suitable for the needs of his tenants and labourers; and at successive periods, owing to the

St John's College

wealth and piety of the lord of the manor, or the zeal of the people, these have been enlarged and renewed; hence we often notice in the same structures examples of all the styles which have been in vogue in this country.

Pre-Norman Period.

In spite, however, of these waves of building enthusiasm which have passed over the country it is possible to discover some of the work of the early masons who were rearing churches before the advent of the Normans. One of the earliest churches in Oxfordshire must have been the minster of St Frideswide's convent. This was almost entirely destroyed on St Brice's Day, 1002 A.D., when the Danes took refuge therein and were burnt with it. It seems to have been of timber, and was rebuilt in stone; and in recent years various remains of this structure have been discovered in the east wall of the north choir aisle of Christ Church Cathedral, formerly the church of St Frideswide's monastery, constructed during the period of transition from Norman to Early English. This Saxon work consists of very rude and early masonry with wide jointing, while outside have been discovered the foundations of two apses which formed the eastern termination of the earlier church. Other pre-Norman work can be seen in the tower of St Michael's church, Oxford, which shows the distinguishing long-and-short work at the angles, and the deeply splayed belfry openings with massive baluster shafts, all typical of the period. Similar signs of Saxon masonry exist in the towers of

St Michael's Church, Oxford

Northleigh and Caversfield, and at Bicester there is a triangular-headed arch of the same early date. Herring-bone work, usually considered a sign of Saxon building, exists in the beautiful church at Bampton. An early window, splayed both ways, is at Swalcliffe, and the churches of Swyncombe, Langford, Broughton Poggs, and Aston Rowant show signs of Saxon work.

Norman Period.

With the advent of William the Conqueror and his followers an era of vigorous church-building commenced. Robert D'Oilly set the example in Oxford, erecting the church of St Mary Magdalen outside the city wall, and possibly that of St Cross at Holywell, the chancel arch of which is early in character. Nothing of his work remains in the former church. The Norman lords who received grants of estates in the county proceeded to build stone churches where formerly wooden structures existed. Very numerous are the examples of this solemn and impressive style, and indeed there are few churches which have no relics of this vigorous architectural period.

Of the early Norman period there are remains at Newnham Murren[1], some windows in the chancel of Sandford St Mary, and the doorways at Ambrosden, Cowley, Wilcote, Stanton Harcourt, Crowmarsh Giffard

[1] This church has been much restored, and injudicious "restoration" often destroys the early features of buildings. The Norman work was done by some of the monks of Bec Abbey in Normandy who were brought over by Milo Crispin, lord of Wallingford, to build churches at North Stoke, Ipsden, and Newnham; but little of their work remains.

and Handborough. The tympanum, or semicircular stone-work between the top of the door and the arch, was often carved with elaborate sculptured devices. At Handborough there is a representation of St Peter seated with a key, his emblem, on one side and the *Agnus Dei* with a scroll on the other, the subject depicting St Peter dictating the Gospel to St Mark. At Kencott Sagittarius is shooting an arrow into the mouth of a dragon; at Salford Sagittarius and Leo are guarding a Maltese cross; at Black Bourton and at South Leigh a simple cross is carved. The semicircular apse at Checkendon and the narrow tower arches at Cassington are also of this early date. The Norman builders tried to introduce apsidal chancels, but the English never liked them; hence many of these apsidal terminations of churches were altered subsequently, and square ended chancels built in their stead.

Of the later Norman period the well-known church at Iffley is one of the finest in the kingdom. The two fine tower arches and the three noble doorways are marvels of the sculptors' art of about the year 1160. Here, as also at Great Barford, Burford, Great Rollright, Asthall, and St Ebbe's and St Peter's-in-the-East, Oxford, are rows of beak-heads, which are supposed to mean the birds of the air in the Parable of the Sower ready to pluck away the good seed sown in the hearts of careless recipients. Somewhat similar to the Iffley door are those of the Chapter House, Christ Church Cathedral, and Woodstock. Immense labour and skill were bestowed on these doorways, which have been often preserved with care in

Iffley Church, West Front

spite of subsequent alterations. We cannot record all the examples, but will mention the richly sculptured tympana at Newton Purcell, Great Rollright, Fritwell, Shirburn, and Bloxham. The Norman builders further ornamented these sculptures with colour, traces of which can still be seen at Shilton, Pyrton, and Brize Norton.

Architects and builders have always been striving after "a more excellent way," seeking for new ideas and new beauties in their art. About the year 1175 they began to improve their methods, seeking to copy nature and not contenting themselves with quaint and curious Romanesque detail. This led to what is known as the transitional Norman style which developed into the Early English. Of this period the best example is Christ Church Cathedral, then the minster church of St Frideswide's monastery. Prior Guymond began the work in Norman times, but as the work progressed the style became lighter and more graceful. The massive pillars of the arcades, alternately round and octagonal, are Norman; the higher clerestory windows transitional. The roof was added in the sixteenth century, probably by Cardinal Wolsey when he began to convert the church into the chapel of his college. He was very ruthless in destroying parts of the minster and of the monastic buildings. The monks required a great church for their services, a cloister wherein they did their work, transcribing manuscripts, writing and illuminating books, a chapter house for their meetings, a dormitory, and a refectory. Wolsey cut down a large portion of the west end of the church, all one side of the cloister, and much else to make room for his college buildings.

Transitional Norman work exists at Cuddesdon, Binsey, Fringford, Enstone, and Broadwell, where there are semi-circular headed doorways. A characteristic feature of Early English work is the pointed arch. There is one at Holton, but the carvings that adorn it are all Norman, showing that it belongs to this period of transition. Other examples can be seen at Kelmscott, Bampton, Swalcliffe, Bucknell, the north porch at Witney, and the west doorway at Shipton-under-Wychwood.

Some curious Norman fonts exist. At Hook Norton the font is carved with figures of Adam and Eve, Sagittarius, and various animals; and others are at Albury, Berwick Salome, Lewknor and other churches. Several Norman lead fonts remain, such as those at Broadwell, Cokethorpe, Warborough, and Dorchester, which has figures of our Lord and the Apostles under round-headed arches. Churches often contain other details, such as piscinae, wherein the vessels used in Holy Communion were cleansed, stoups for holy water near the entrance, sedilia (seats for the clergy), aumbrys or cupboards wherein the treasures of the church were kept. Many of them are of Norman date.

Early English Period.

At length the English builders found their way to the new style of Gothic architecture, boldly attempted the pointed arch, and sought in nature models for their sculpture. We have numerous examples of this style in our county. The beautiful spire of the Cathedral, that

Christ Church Cathedral

at Witney together with most of the church (1200 A.D.), the Chapter House at Oxford (all the eastern portion) with its delicate lancet windows, the chancel and transepts of Stanton Harcourt with rows of lancets, are all in this style. The thirteenth century was the "golden age of churchmen," and foremost amongst them was Bishop Robert Grosseteste of Lincoln, in whose diocese Oxford then was. He was a great builder and his work remains in the church and the chapel of the Prebendal House at Thame, where there is a fine east window of three lights, delicate detached and clustered shafts, and rich capitals. The church built by him, with its fine Early English arcades, was much added to in the fourteenth century; the accounts of this work are still in existence, and a copy is in the writer's possession. Another great church builder of the time was Richard, Earl of Cornwall, brother to Henry III, and king of the Roman Empire. His wealth was enormous and he loved to spend it in building churches. Many of the Oxfordshire churches owe their beauty to his bounty. His badges, the lion rampant crowned and the spread eagle, are found in many churches in stained glass and on tiles, as at Ipsden. Among other examples of the work of the period may be mentioned Tackley church with its eastern lancets, Wigginton, Kidlington, and Bucknell. North Stoke has an Early English spire, though the upper story is modern, and an ornate chancel of this period. This church must have been in decay in the fourteenth century when it was given to the convent of Bromhale, Berkshire, as the Bishop of Lincoln ordered the Prioress to restore it. Hence there is much Decorated work of the date 1320. We find other examples of Early English work in the south aisle of Swalcliffe (1250 A.D.), and in the doorways of Great Haseley, Great Milton, and Charlbury.

Decorated Period.

Most of the triumphs of art in the ecclesiastical architecture in the county belong to the Decorated period. We have already noticed the affection which some of the Bishops of Lincoln retained for this distant corner of their huge diocese. This was shown by the building of the beautiful church at Banbury, mainly erected at the end of the thirteenth century. It cannot be seen now, as it was entirely pulled down in 1790 in order to save the expense of restoring it. One of the most interesting churches in the county is Dorchester Abbey Church, once the seat of the bishop, before it was transferred to Lincoln. In 1140 Bishop Alexander of Lincoln established here a priory of Austin Canons, and the minster testifies to their skill as builders. The present building was soon commenced by them, but its chiefest glory is the work of the fourteenth century, the choir, south choir aisle, and south nave aisle. "The extreme splendour of the arches on each side of the choir must strike every one," wrote Professor Freeman. "They are the finest I know." He says again, "There is probably no existing building which shows a greater number of singularities crowded together in a small compass than the eastern bay." The great east window is most remarkable, and still more the extraordinary Jesse window on the north. The Patriarch lies below and from his side springs his genealogical tree, carried up in the stone mullions and in stained glass. This church would require a volume for its full description. Going back to Oxford we there find the builders very busy. They were building Merton College Chapel, inserting windows in the Cathedral, and restoring many of the churches of the town. St Mary Magdalen's church, originally Norman, was rebuilt at this period. Owing to its position between two streets it could not be extended eastward or westward; hence all its expansion has been in the form of aisles on the north and south. The masons of this period were very fond of light and inserted many beautiful windows with elaborate tracery in many churches, such as Great Haseley, Great Milton, Bampton, Kidlington, and in the chancel of Piddington. One peculiar feature of the Oxfordshire churches in the northern part of the county and adjacent districts is the fringe or foliated canopy to the containing arches of the windows, such as at Broughton, Bampton, and Broadwell.

Of later Decorated work, called sometimes curvilinear style, there is the beautiful Latin Chapel in the Cathedral, which is an architectural gem, and also the fine churches of Cropredy, Broughton, and Chinnor. The chancels of Merton, Chalgrove, Garsington, and Stanton St John, are almost entirely of this period, and the north transept at Witney and north aisle at Ducklington with its beautiful canopied tombs. The nave, tower, and spire of Adderbury church, the spire of St Mary's, Oxford, with its

St Mary's Church, Oxford

profusion of ball-flower ornament, and the north aisle of Swalcliffe, said to have been designed by William of Wykeham, the great architect and bishop, the inventor of the Perpendicular architecture, before he had conceived that peculiar style, are all of this period. Many of the churches at that time were adorned with mural paintings, and attained to the height of their magnificence. They were a blaze of colour. But it would take too long a space to record all that time has spared.

Perpendicular Style.

Oxfordshire, as we have seen, is closely connected with the originator of this style, which is peculiar to England, the great architect, Bishop William of Wykeham, who remodelled Winchester Cathedral. As the founder of Winchester School and its associated college at Oxford, New College, he introduced his architectural ideas when building the chapel towards the end of the fourteenth century.

He had other connections with the shire, having purchased Broughton Castle for his nephew, Sir Thomas Wykeham, upon which he exercised his skill in building, and designed the chancel of Adderbury church. The chapel of Magdalen College, the churches of Handborough, Chipping Norton, and Ewelme, and the font at Burford are also examples of the Perpendicular style. The church of Minster Lovell, cruciform and with central tower, was built by Lord Lovell in the time of Henry VI, with its fine founder's tomb and font, and Richard Quatremain

Magdalen College

and Sibylla his wife in 1449 built the curious Rycote chapel, which remains entirely unrestored.

Burford Church

At this time the clothiers and wool-merchants were very prosperous in the Cotswold district and spent their wealth freely in the building or re-building of churches. Thus, the noble church of Burford was almost entirely re-edified. Especially noticeable are the upper part of the tower, the spire, the south porch with fan-traceried roof, and the font with carvings of saints and the Crucifixion.

The monastic buildings have fared badly. They have

Godstow Nunnery

left few remains. In the county there were five Benedictine houses, three Cistercian, seven houses of the Austin Rule, one Gilbertine priory, two alien priories, Coggs and Minster Lovell, a house of the Templars at Sandford and of the Hospitallers at Clanfield. Besides these there were five monastic colleges and several hospitals. Though not very rich and powerful abbeys and priories, they were fairly numerous. Godstow and Shipton-under-Wychwood are ruins. Some were converted into private houses, or made into farms, or entirely pulled down in order to make way for new structures. Studley and Bicester, Wroxton and Elvenden betray a few traces of their origin. The fine medieval hospital at Ewelme remains.

Renaissance.

During the sixteenth century English architecture began to be influenced by the Renaissance, as the revival of art in Italy was called. We find classical details gradually creeping into use in the Gothic designs during the Tudor, Elizabethan and Jacobean periods, but the great change was due to the genius of Inigo Jones who first mastered the spirit of the Renaissance architecture, and adapted it to English needs. This apparently entirely new idea of architecture was, as is so often the case with new ideas, only a return to a much older source of inspiration. For our Norman style had been developed through the French Romanesque and Byzantine styles, from the older Roman and Greek architecture.

The tower of the Five Orders in the old University Schools, Oxford, illustrates the main features of Renaissance architecture. The five orders are Tuscan, Doric, Ionic, Corinthian, and Composite.

It is curious, however, that at Oxford there was an after glow of Gothic architecture. In spite of the changes wrought elsewhere the masons at Oxford clung to their

Tower of the Five Orders

Christ Church: Staircase leading to the Hall

traditional methods, and wrought as their fathers had done before them. This is noticeable in the buildings of Wadham, in the fan-traceried roof of the entrance to the hall at Christ Church and elsewhere.

The blending of Classic and Gothic forms is well seen in the chapel of Brasenose College (1666), attributed (but wrongly) to Wren, who with his pupil Hawksmoor designed Queen's College buildings and the chapel of Trinity College. The Bodleian Library, built in 1602–1636, is a fine example of Renaissance style described by Casaubon as "a work rather for a king than a private man." Many churches contain Renaissance details, porches and monuments. The noble monument to Sir Lawrence Tanfield in Burford church (1626) is a notable example.

17. Architecture—(*b*) Military. Castles.

We have already recorded many of the great earthworks and lines of defence in Oxfordshire which were thrown up in prehistoric times, and denominated "castles," but in this section we are concerned only with military strongholds built of stone. Some few remain, but most of those erected in troublous times have disappeared.

The Normans when they came built many castles to overawe the conquered English folk. Their earliest castles were often simply fortified mounds with timber forts, but soon these wooden walls were replaced by stone. Robert D'Oilly was ordered to build such a castle

Oxford Castle

at Oxford, where the great square keep—" four square to every wind that blew"—still watches over the city. The walls of these castles were very thick, and often 50 feet high. A deep well supplied the inhabitants with water, and the keep was divided by several floors, access to which was gained by spiral stone steps laid in the thickness of the wall. A moat surrounded the whole castle, crossed by a drawbridge, protected on the side remote from the castle by a barbican. High walls with an embattled parapet surrounded the lower court, or outer bailey, which was entered by a gate defended by strong towers. In the lower court were the stables. Another strong gateway, flanked by towers, protected the inner court, where stood the keep, chapel, and barracks.

Oxford Castle was built on a high artificial mound erected in Saxon times whereon stood the less formidable-looking fortified dwelling of English and Danish rulers. It has played a prominent part in the history of the city. Its strength and position made it of great importance, as it ranked with Windsor and Wallingford as the three principal guarding-places of the valley of the Upper Thames. It was not often the residence of the monarchs, who usually stayed at their neighbouring palace of Beaumont, or at Woodstock. During the Civil War of Stephen's reign it was held by the Empress Maud, who took it from Robert D'Oilly, nephew of the founder, and it endured a siege of eight weeks, when the food began to fail, but the Empress contrived to escape in the dramatic manner already described. It was doubtless within the castle walls that the Council was held in 1133 which settled the terms of peace, and numerous Parliaments assembled here. In the wars of the Barons in John's reign the castle was a valuable stronghold of the king and dominated the country-side. It was the headquarters of those barons who supported the young King Henry III, while Lewis the Dauphin and his followers took up their position at Cambridge. Little is known of the plan and building of Oxford Castle. All that is left is the keep, the piers of a Norman crypt, and a vaulted chamber containing a well.

In Oxfordshire, as in other parts of England during the troublous times of Stephen's reign, several castles were erected by nobles and landowners which became dens of robbers for the pillaging of the country. Such castles, called *adulterine*, were raised at Ardley (by the Earl of Chester), Swerford, Somerton, Chipping Norton, Mixbury, Deddington, and Bampton.

When Henry II came to the throne he ordered these castles to be destroyed and razed to the ground. Ardley, Swerford, and Somerton were pulled down, but the rest were spared. Swerford was probably built by Robert D'Oilly junior, nephew of the Conqueror's ally. "Castle Hill" still marks its site with its great mounds. The moat and earthworks of Ardley Castle remain, and Somerton Castle was built by one of the Arsic family of Coggs.

Bishop Alexander of Lincoln built a fine castle at Banbury in 1135, which has a notable history. It was the Oxfordshire residence of the Bishops of that see until the time of Edward VI. It played a great part in the Civil War of the seventeenth century, and was of considerable strength. It was held at first by the governor Nathaniel Fiennes for the Parliament, but after the battle of Edgehill he yielded it to the Royalists without a struggle, and all through the war it was gallantly defended by the King's soldiers and proved a thorn in the side of the Parliamentarians. The inhabitants of the town and district favoured the latter, and were constantly pillaged to supply food for the garrisons of Oxford and Banbury. It was vigorously besieged in 1644 by Colonel John Fiennes and as gallantly defended by Sir William Compton, who returned this answer to a summons to surrender, "We keep this castle for his Majesty, and as long as one man is left alive in it we will you not to expect to have it delivered."

The old church was the headquarters of the besiegers, who planted their cannon in the churchyard. The weary siege lasted thirteen weeks. Then a vigorous attack was made, many soldiers being slain, but the besiegers were driven back. The garrison was in sore straits as they had only two horses left for food. Two years later the attack was renewed when the royal cause had failed, and the castle being delivered up, was slighted and destroyed by the Roundheads.

A castle was built by Gerard de Camville at Middleton Stoney in the time of Stephen, whose cause he supported against the Empress Maud. His son Richard there mustered his adherents to accompany him to the Holy Land on the great crusade under Richard, the Lion Heart, but he never returned to see his home again. His son Gerard succeeded and held the castle for Richard's enemy John. On King Richard's return he was only allowed to retain his castle after paying 2000 marks. Leland in the time of Henry VIII described its ruins, and says the castle stood near the church. "Some pieces of the walls of it yet a little appear, but almost the whole of it is overgrown with bushes." Some mounds still mark its site.

Another early castle existed at Chipping Norton, built by the Fitz-Alans. Erected in Stephen's reign it escaped destruction in the time of Henry II, when some other castles were pulled down. The "Castle Banks" near the church are all that remain of it. Roger d'Ivry built a castle at Mixbury of considerable strength and so beautiful that his Norman neighbours called it Beaumont, or *de Bello Monte*. No stones of it are standing, but it is possible to follow the outlines of the strong building that once stood there, and mark the position of the keep and the inner and outer bailey-courts, while the moats show where the walls of the fortress stood.

It is not known when the strong castle of Deddington was raised, but it was in existence in the time of Edward II, and was then held by Aymer de Valence, Earl of Pembroke, who also had licence from the king to crenellate his house at Bampton in 1315, that is, to erect a castle. Deddington had a notorious prisoner, Piers Gaveston, the worthless favourite of Edward II, who offended all the chief nobles of the realm by his pretensions, his manners, and by the offensive names he gave to them. The owner of Deddington and Bampton he called "Joseph the Jew," the Earl of Warwick he nicknamed the "Black Dog." At length they seized Gaveston and held him prisoner at Deddington, ere they carried him off to the scaffold near Warwick. Some green mounds cover the remains of the once strong fortress.

In the reign of the Edwards a new style of castle-building arose. No longer did the towering keep dominate the fortress. The Edwardian castle was based upon a square or oblong plan with towers at each corner and high curtain walls between them and a strong entrance gate in the centre of one of the sides. Anthony Wood visited Bampton in 1664 and made a sketch of the castle. This drawing is still in existence and shows that the fortress corresponded with the Edwardian type. It was of a quadrangular form, with a moat round it, and had towers at each corner, and a gatehouse of tower-like character on the west and east sides. At that time the whole of the western front was standing. The ruins stand a short distance westward of the church, from which it is separated by a brook. A gateway and a fragment of wall furnished with loopholes and battlements still remain. The groined roof of the upper chambers, the spiral stone staircase, and the niches in the walls with narrow slits for the discharge of arrows are interesting features of this once strong castle. There is a field near the castle which was formerly used as a tilting ground, where tournaments and the display of ancient chivalric exercises took place. It is also surrounded by a moat, and in the hollow ground formed by the crumbling sides of the moat is a holy well whose healing waters cured all kinds of diseases. It was associated with the Blessed Virgin Mary, and was much frequented until the beginning of the last century. Aymer de Valence, the bitter enemy of Piers Gaveston, built this castle and also owned Deddington, fought at Bannockburn, and was killed at a knightly tournament. His splendid tomb is in Westminster Abbey. It is recorded that on one occasion he took part in a joust at Witney with Humphrey Bohun, Earl of Hereford. The castle passed by the marriage of his daughter to the Earls of Shrewsbury and remained in their hands until the beginning of the eighteenth century.

Another Edwardian castle was built near Henley-on-Thames, known as Greys Court, erected in 1350 by John de Grey, whose family gained large possessions in the county, their memory being preserved by the name of the village Rotherfield Greys. Four of the towers are still standing and there are the remains of a moat. Within the walls of this castle a noble Elizabethan house was erected, now the residence of the Stapleton family.

Shirburn Castle, also of the Edwardian type, was built originally by Warine de Lisle by royal licence in 1377. A previous castle erected by Robert D'Oilly existed here, and was in the hands of King Stephen during the civil war; but the Empress captured one of the King's supporters, William Martel, and agreed to release him on condition that the castle was delivered up to her. It played a prominent part in the troubles of the reign of Edward II. Hither came Thomas, Earl of Lancaster, and the other Barons to conspire to overthrow the power of the Despensers, the King's favourites, who proved too strong for the nobles, and the owner of the castle, Warine de Lisle, was beheaded. It was his

Broughton Castle

grandson who began to build the present castle, which was not finished a century later. A wide moat surrounds it, spanned by a drawbridge, and a portcullis still exists. It is similar in plan to the Edwardian castles already described, but has been somewhat modernised. It has been held by many distinguished families, and was besieged during the Civil War in the seventeenth century, and gallantly held by the wife of the owner for a long period until the royal cause seemed hopeless and the full force of Fairfax's army was brought against her. It is now the residence of the Earl of Macclesfield.

Banqueting Hall, Broughton Castle

Broughton Castle at the present day belongs to the type of fortified mansions built at a time when the need of extensive castellated works and massive protecting walls had passed away, and the country had become settled. Recent discoveries have shown, however, that a considerable part of the castle was erected in the fourteenth century (1301–1307) by the Broughton family, who derived their name from the place. This earlier portion of the castle includes the hall, its date having been determined by the discovery of some fine Decorated windows. It would take much space to describe the subsequent alterations which have almost converted the appearance of the castle into that of an Elizabethan mansion. The guarding gatehouse, with the exception of the upper storey, the chapel and priest's room, the hall and some passages and rooms, are all fourteenth century work. The castle was purchased by William of Wykeham, the great architect, and given to his nephew Sir Thomas Wykeham, who considerably added to it. Then came the Lords Saye and Sele, who still own the castle, William, second Baron Saye and Sele, having married Margaret, great niece and heiress of Sir Thomas Wykeham. This distinguished family brought the castle into close touch with the annals of England, and the Elizabethan portion of the house was added by them. James I visited it on two occasions. It became the focus of resolved resistance to Charles I, the Great Rebellion having been planned in the little room at the top of the house where Pym, Hampden, Brooke, and Lord Saye and Sele (known as "Old Subtlety") used to meet. The Saye and Sele Bluecoats fought at Edgehill. The castle was besieged by the Royalists. An earthwork was thrown up in the park for a battery, and the defenders hung bales of wool over the battlements to break the cannonading. The castle surrendered and was plundered, but its walls were not injured. Many relics of the Civil War period are preserved in the castle. "Old Subtlety" became a staunch Royalist.

In addition to these castles the fortified manor houses in the county may be mentioned, which proved themselves formidable garrisons in the Civil War period. Such were the royal manor of Woodstock, which made a gallant defence; Fawley and Phyllis Court near Henley; the Blounts' house at Mapledurham; and many others which on account of their formidable defences might claim to be ranked as castles.

18. Architecture—(*c*) Domestic. Famous Mansions, Manor Houses and Cottages.

Oxfordshire can boast of many noble and stately houses, and even the humble cottages of the countryside are worthy specimens of domestic architecture. Great Tew, which lies amidst steep, well-timbered hills in mid-Oxfordshire, has the credit of being the prettiest village in the shire. All the cottages are built of a local stone which has turned to a grey colour or rich ochre, and they are either steeply thatched or roofed with thin slabs of the same yellowish grey stone. The diamond-paned windows often have stone mullions with dripstones over them. No one cottage repeats another. Nowhere do we find slate or red brick. It is all very harmonious and beautiful.

Great Tew Village

The story of the growth of English domestic architecture is well illustrated by the manor houses and mansions of Oxfordshire. First there are the thirteenth century manor houses of Cottisford and Coggs. The former has been much altered and is now much smaller than it was originally, but it retains some fine Early English windows. The village of Coggs is very remarkable for its church, vicarage, barn, and manor house, which form a striking group of old buildings. The manor belonged to Odo, Bishop of Bayeux, after the Conquest. There was a priory here founded by William de Arsic and attached to Fécamp Abbey, Normandy, and some part of the monks' buildings is incorporated in the vicarage. The church shows signs of French influence. The manor house retains its Early English windows, but was mostly rebuilt by Earl Downe in the sixteenth century.

Bishop King's Palace was built in 1546 for the first Bishop of Oxford and last Abbot of Osney. The front was rebuilt in 1628. It has some richly decorated ceilings.

Of the fortified structures reared when "every man's house was his castle," some examples have been given. When the wars were ended and the country more secure, men began to build more comfortable dwelling-places. Tudor architecture has the great charm of homeliness and severe beauty, and has produced some of the best types of houses in England. The plan of a Tudor house is very much like that of an Oxford or Cambridge college. There is an entrance-gate with porter's lodge or guard-room for retainers. This opened into a square court, on the far side of which we see the

Bishop King's Palace, Oxford

main entrance, which admits to a passage. On the right is the hall, separated from the passage by a screen; on the left are the buttery and kitchens; over the screen is the minstrels' gallery. The hall has a dais, or raised platform at the end remote from the screen, where the lord and his family dined, the servants taking their meals in the lower part of the hall on tables set on trestles. On the dais side there were doors and staircases leading to the withdrawing room and private apartments of the family. Based on such plans was the old house of Minster Lovell, now in ruins, remarkable as being a purely domestic building with no military feature, save a moat. The hall was a stately building of great height, lighted by four fifteenth century windows. History books tell us the rhyme:—

> "The Cat, the Rat, and Lovell the dog
> Rule all England under the Hog,"

referring to John, Lord Lovell, who took office under Richard III and was hated by the Lancastrians, together with Catesby (the Cat), and Ratcliffe (the Rat). The son of this Lovell was in hiding in a vault at the house, his wants being attended to by a faithful old servant; but she died suddenly, and her master was starved to death.

Stanton Harcourt is a good example of an Early Tudor house as far as can be judged from the scanty remains. Henry VIII introduced foreign craftsmen into England who altered our style and greatly increased the ornamental details. Elizabethan houses have a profusion of ornament.

The long gallery became a feature of the house; it was on the first floor, and elegant staircases were introduced to lead to it. In later days came in the fashion to build in the Italian style, and many beautiful old houses were pulled down in order to make room for these foreign innovations.

Wroxton Abbey

Broughton Castle has much Elizabethan work. Wroxton Abbey is mainly Jacobean. There are a few thirteenth century arches remaining of the old monastic building. The present house is a very noble dwelling, full of historical portraits and tapestries, and beautiful in every respect. Studley Priory was also built in the time of

James I, and here too we can discover some traces of the monastic buildings. Asthall Manor is Elizabethan: Chastleton was built between the years 1603 and 1610. Robert Catesby of Gunpowder Plot fame once owned the manor and sold it to Walter Jones, whose family still holds it.

Some houses are built on a plan shaped after the letter E; some think that this was in compliment to Queen Elizabeth, but that is a mistake. It was a development of preceding plans. Thus we have the original plan of a hall. Then the house took to itself wings in order to provide additional rooms for the family

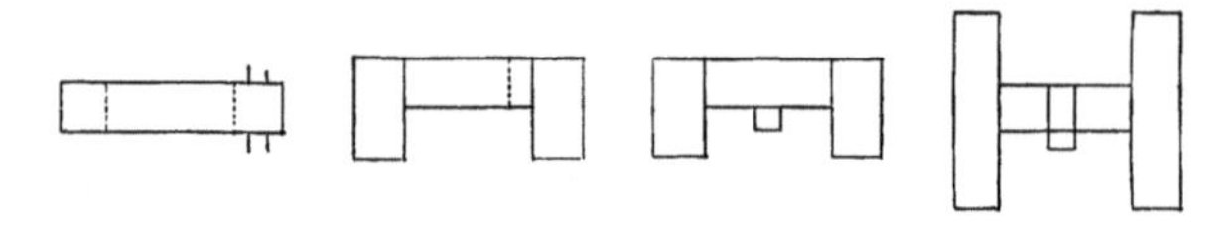

and servants and the porch became a more elaborate structure. Thus we get an E-shaped house. There are some H-shaped houses, and these are taken to stand for Henry VIII. But this is only a further development of plan. The cross piece is the original hall, and the wings on each side are extended both ways in order to provide increased accommodation for the family and guests. Fritwell is E-shaped; it is a very beautiful Jacobean house erected in 1619. Shutford manor house is earlier, a lofty, gloomy house built by the Wykehams of Broughton. Mapledurham, the house of the Blounts, was mainly built in the time of Elizabeth, when the beautiful front was added to an old irregular half-timbered house of the fifteenth

The Octagon House, Oxford

century. It has beautiful oriel windows, a central portion with two deep wings, and high towering chimneys of red brick, of which the house is built.

There are other good examples of domestic architecture in Hardwicke, Barton, Cornbury, Rousham, Gaunt House, Cote, Shipton Court, and Stonor. But these are sufficient to show the treasures which the county contains.

Blenheim Palace, presented to the first Duke of Marlborough by the nation, is a building of great magnificence and cost £300,000. It was built by the playwright and architect Sir John Vanbrugh in 1704. Other later mansions are Nuneham, Heythrop, Tew Park, Middleton Stoney (1755), Tusmore (1766), and Watlington.

An illustration is given overleaf of a house called the Octagon House, said to be the oldest in Oxford. The doorway appears to be of the fifteenth century.

Some mention of cottages has already been made. These are especially beautiful at Burford and in the Cotswold country. We see there the greatest pains bestowed upon the details of these buildings, on the doorways, windows, chimneys, and roofs. There is nothing hurried or slovenly about them. They differ greatly from modern cottages, which are built cheaply with glaring bricks, ugly slates, and the cheapest possible door and window frames imported from abroad. Nothing can be more hideous than rows of modern cottages, or more beautiful than the old Oxfordshire cottages of the Cotswold country.

19. Architecture—(*d*) Colleges.

The history of the founding of the University of Oxford abounds in interest and needs careful study. Popular tradition points to King Alfred as its first founder, and in his time there may have been schools at Oxford in Saxon days. Oxford owes much to the University of Paris, where there was an outburst of great intellectual activity at the end of the eleventh century, and Abelard grew up to be the greatest of its teachers. Paris scholars formed themselves into a definite society about the year 1150. English scholars often went there to seek learning, and sometimes Frenchmen came to Oxford for a like purpose, as in 1117 we hear of a Doctor of Caen, in Normandy, becoming a master of Oxford. As all scholars used the Latin tongue there was much more intercourse between those of various countries than in later times. Towards the close of the twelfth century a great migration took place of students from Paris to Oxford, and soon we find 3000 students at the University. They were lodged in halls, hostels, and boarding houses, but the colleges did not yet exist. University College, originally Great University Hall, was founded in 1249 by William de Durham. It was not really a college until later, and no part of the present building is earlier than 1634. The real founder of colleges was Walter de Merton who in 1263 made over his estates to a community of scholars who were to study at a University. This was the origin of Merton Hall or College, and after his time colleges sprang up and became the homes of learning, religion, and fellowship which they now are.

The usual plan of a college, as already stated, is somewhat similar to that of a large Tudor house. It required a great hall with kitchen, buttery, etc., a chapel, and rooms for tutors, fellows, and students. The buildings were grouped round one or more large quadrangles, and some of them retain the beauty of their Gothic architecture. Many have been rebuilt in later times.

It is curious to note that at Oxford Gothic traditions of building lingered on long after they had died out elsewhere. The builders and masons were accustomed to build in Gothic style and continued to work in that style regardless of the changing fashions elsewhere. Thus in the seventeenth century they built Wadham College, a fine late Gothic college, when the Renaissance style was being followed everywhere else.

A walk through the streets of Oxford is one of the most delightful that can be enjoyed. Starting from Carfax, a corruption of *Quatrevois* or four ways, we pass along St Aldate's Street, named after the church dedicated to the Saxon Saint, and on the right is Pembroke College, formerly Broadgates Hall. The present college was founded in 1624 and named after William Herbert, Earl of Pembroke, Chancellor of the University in 1616. Only the refectory of the older buildings remains. The front quadrangle was built in the seventeenth century, the chapel in 1732 and the new hall in 1848. Dr Johnson became a student here in 1728. On the left of St Aldate's is Christ Church, the most magnificent college in Oxford,

The Great Quadrangle, Christ Church

founded by Cardinal Wolsey in 1525 and at first called Cardinal's College. After his fall King Henry VIII completed the foundation, though on a smaller scale than the Cardinal intended. We enter the great quadrangle ("Tom Quad"), which takes its name from the famous Bell, Great Tom, that hangs in the gateway. The college buildings share the remains of the monastery of St Frideswide, including the minster (now the cathedral, which was shorn of part of its length by the college buildings), the chapter house and the refectory. The tower that carries the great Bell was built by Sir Christopher Wren in 1682. The beautiful fan-tracery in the roof of the staircase leading to the hall looks like Perpendicular work, but was made by one Smith of London in 1640. The hall is a noble structure and its walls are hung with the portraits of the famous sons of Christ Church. The new buildings erected in 1862 in the style of "Venetian Gothic" are an illustration of how *not* to build. There is another quadrangle called Peckwater, named after an old hostel kept by Ralf Peckwater in the thirteenth century. The present buildings are Palladian, so named after Palladio, the Italian architect who flourished in the middle of the sixteenth century and set this fashion of building in this country. A small quadrangle adjoins called Canterbury, so named after a college of that name founded by Simon Islip, Archbishop of Canterbury in 1363. Together with Peckwater Inn it was united with Christ Church by Henry VIII. The present buildings were erected at the close of the eighteenth century.

Leaving Christ Church by this Canterbury Gate we see Oriel College, founded by Adam de Brome, almoner of Edward II, and originally called St Mary's College. The king, however, acquired the patronage and credit of founding the college. Its name is somewhat puzzling and is probably derived from an old mansion called "The Oriole" which formerly occupied the site. Its old

Oriel College

buildings have entirely disappeared, the present front quadrangle having been erected in 1620. It is an interesting example of "Oxford Gothic," like Wadham, retaining many characteristics of earlier traditional work. Noticeable are the mullioned windows, hooded by labels, the handsome shaped gables, the hall on the left with its louvre, the fine porch with its frieze "Regnante Carolo," the statue of the Virgin, and those of Edward II and Charles I. The old St Mary's Hall is now incorporated with the college. The inner "quad" (Oxford men always call these courts quads, not quadrangles) shows in the library, erected in 1788, the debased architecture of Wyatt, the architect who did so much mischief in restoring cathedrals and destroying the beautiful work of

Merton College Library

earlier ages. Amongst its great men was Sir Walter Ralegh, and Pusey, Keble, Dean Church, Tom Hughes, and Cecil Rhodes were also members of the college. In 1911 the college was extended into the High Street, the cost being defrayed out of the Rhodes bequest.

Near at hand is Corpus Christi College, founded in 1516 by Richard Foxe, Bishop of Winchester, prelate, statesman, architect, soldier, and diplomatist. The front quadrangle is his work and also the chapel, though it has been much mutilated by subsequent alterations. Proceeding along Merton Street we come to Merton College, the first real example of the collegiate system. The parish church of St John Baptist became its chapel. The oldest part of the present buildings is the muniment room, which may date back to the original tenements purchased by the founder in 1263. The hall was restored and ruined by Wyatt, and restored in 1878. The library, with its chained books, is one of the most interesting medieval rooms in England; it was erected in 1345 and the beautiful quad on the south was completed in 1610. The entrance gateway was built in 1416 and has statues of Henry III and Walter de Merton, and a sculptured tablet representing St John preaching in the wilderness.

At the eastern end of the High Street are the grand buildings of Magdalen College, the most beautiful piece of architecture in Oxford. It was built by William de Waynflete, once Master of Eton in the time of Henry VI, in 1458. The tower was completed in 1507 and together with the cloisters form a remarkable group of great beauty; its sons have been noted for their learning, and include two cardinals, four archbishops, forty bishops, and such eminent men as Sir Thomas Bodley, founder of the Bodleian Library, Dean Colet, Addison, Gibbon, Lyly, Hampden, and Camden.

Next we notice University College, founded as Great University Hall in 1249. The oldest part of the present buildings is the west side of the first quad, begun in 1634.

Magdalen College Tower

A legend connects the college with King Alfred, but this is entirely mythical. Amongst its alumni are Robert Dudley, Earl of Leicester, Richard Fleming, founder of Lincoln College, several bishops, Shelley, and many other distinguished men. Across the way stands Queen's College, remarkable for its Italian-looking front, designed

Queen's College

by Hawksmoor, a pupil of Wren. It was founded by Robert de Eglesfield, chaplain to Queen Philippa, after whom it is named, in 1340. Several old customs are preserved at Queen's, such as the bringing in of the Boar's Head on Christmas Day, and the giving of a needle and thread on New Year's Day, with the words "*Aiguille et fil*—Take this and be thrifty." The French words mean needle and thread, and are supposed to sound like "Eglesfield," and thus perpetuate the memory of the good founder's name.

Not far away is All Souls' College, the splendid foundation of Archbishop Chichele in 1437, a college of Fellows, not of undergraduates. Near it rise the lofty spire of St Mary's Church, the dome of the Radcliffe Library, the University buildings and Bodleian Library, and Brasenose College, founded in 1509, with its brazen nose above the gate. The name is usually derived from Brasen-huis, a brew-house, which formerly occupied its site. Near by in Turl Street is Exeter College, founded in 1314 by Walter de Stapledon. Little old work remains, only the gatehouse and the Rector's lodgings. Here also is Lincoln College, founded in 1427 by Richard Fleming, Bishop of London, "to defend the mysteries of the sacred page," and to oppose the doctrines of Wycliffe; and Jesus College, a Welsh foundation, the first Post-Reformation College in Oxford. Hugh Price was the founder in 1571 with the patronage of Queen Elizabeth, whose fine portrait by Zucchero hangs in the hall. The oldest part is the front facing Turl Street. The chapel is "Oxford Gothic" of the date 1636.

Across Broad Street stands Trinity College, founded on the site of the old Durham College of the thirteenth century, by Sir Thomas Pope, one of Henry VIII's most trusted officials. Wren designed the garden quadrangle, a fine piece of work. On the west of the front quad can be detected some original building. Next to it stands

The Twin Towers, All Souls'

The Radcliffe Library

Balliol College, noted for its scholarship and its antiquity. Sir John de Balliol was a powerful baron in the time of Henry III who joined Simon de Montfort's rebellion. He was captured and ordered to be scourged at Durham Cathedral. He escaped the penalty by promising to found a college for poor Durham scholars. Dying before accomplishing his intention, he begged his widow Devorgilla

New College

to build the college. She carried out his wishes and created a foundation which subsequently developed into Balliol College. Very little of the old building remains, only the reading room and library, which are early fifteenth century work. Modern buildings have not improved the college, but it remains a home of light and learning, and numbers amongst its alumni a crowd of great names, including Evelyn, Lockhart, Southey, Matthew Arnold, Swinburne, Manning, and T. H. Green.

New College is a bright contrast to Balliol. It has preserved its old buildings intact erected by William of

Mansfield College

Wykeham just before 1400 A.D., though the roof of the hall has been raised and a third storey added to the front quad. William of Wykeham intended that it should be allied with his other foundation, Winchester School, in order to complete the education of the students. St John's

College was founded by Archbishop Chichele, the pious founder of All Souls'. Its original name was St Bernard's College, and it was connected with the Cistercian order of monks. This was swept away by Henry VIII, and in Mary's reign Sir Thomas White, twice Lord Mayor of London, refounded the college and designated it

Somerville College

St John's College. Archbishop Laud's munificence enabled Inigo Jones to complete the second quad, which opens into one of the most beautiful gardens in Oxford. Wadham College was founded in 1612 by Dorothy Wadham in accordance with her husband's wishes. The buildings are as she left them, a fine example of "Oxford Gothic," retaining traces of the Somerset origin of its masons. Worcester College was formerly Gloucester Hall, and Hertford College in living memory Magdalen Hall. Keble College is modern, reared in memory of John Keble, the author of the *Christian Year*. Mansfield is also modern and was built for the benefit of Nonconformists. Nor are women students forgotten in modern Oxford. Somerville Hall was opened in 1879 and afterwards enlarged, and there are also Lady Margaret's, St Hugh's, and St Hilda's Halls.

We have traced the history of these Oxford colleges which form part of the great University, the *alma mater* of learning and of learned sons. She has played a great part in the history of England, and by wise effort and expansion may live to extend her influence still wider and to confer increased benefits upon the nation.

20. Communications—Past and Present. Roads. Railways. Canals.

The oldest roads in Oxfordshire are the trackways which were made in prehistoric times, connecting early settlements, and often guarded by camps and earthworks. Such a trackway runs from Rollright Stones northward along the top of the hills separating the territories of the Dobuni and the Carnabii as far as Nadbury Camp. Another runs from the same stones by Hook Norton Camp and through Tadmarton Camp, through Banbury, connecting with Banbury Lane a long trackway leading into Northamptonshire. Along these ancient trackways about Rollright and Tadmarton drovers could travel, until the middle of the last century, more than one hundred miles without passing through a tollbar. The Saltway, another ancient trackway, which still exists under that name at Banbury, ran by the foot of Crouch Hill in a south-easterly direction towards London. The

The Saltway near Banbury

Portway crossed the Banbury Lane from north to south, on the east of the district adjoining the Coritani. It entered Oxfordshire at Souldern, proceeding south to Kirtlington and going on to Port Meadow at Oxford. It still remains a direct and ancient trackway. The Icknield Way, running from the east coast to Bath, passes through Oxfordshire. It enters the county near

Chinnor, and then divides into an upper and a lower road. Passing by Aston Rowant and then east of Watlington it skirts the Swyncombe Downs and proceeds in a south-westerly direction to Goring, crossing the Thames to Streatley in Berkshire.

The Romans often made use of these old trackways, improving them and paving them. The chief authority for Roman roads in Britain is the *Antonine Itinerary*, a road-book of the Roman Empire, probably written in the reign of Antoninus Pius (138 to 161 A.D.) but largely added to in later times. But it omits to record the great Roman road called Akeman Street, a branch of the Watling Street, that ran through Tring and Aylesbury to Cirencester and Bath. It was probably a later Roman road and therefore not included in the *Itinerary*. It enters the county from the east near Blackthorn, not far from Bicester, proceeds to Kirtlington across the Cherwell, through Wychwood Forest to Asthall, where it crosses the Windrush, and then goes up the hillside across the Cheltenham road, and so into Gloucestershire to Cirencester.

Another Roman road has been traced from Alchester to Dorchester running across Otmoor. It probably crossed the Thames and passed on to Calleva or Silchester.

The building and repair of bridges and the maintenance of roads in medieval times were considered religious acts. Guilds existed for this purpose, and the monks kept the roads in repair on their properties. They fell into sad neglect after the destruction of monasteries and guilds. Arthur Young, who travelled about England observing the state of the country, wrote in 1740 that he remembered the roads when they were in a condition formidable to the bones of all who travelled on wheels. The two great "turnpikes"—as they were called—which crossed the county by Witney and Chipping Norton, by Henley and Wycombe, were repaired in some places with stones as large as they could be brought from

Whitchurch Toll Gate

the quarry, and when broken, left as they were. At that period the cross roads were passable, but with real danger. There was always a wide space on each side of the road, so that when the road itself was very bad vehicles could be driven along these side ways. Great improvements were made at the beginning of the nineteenth century. This was caused by the passing of the

Turnpike Act in 1763, which ordered the levying of tolls for the repair of the roads. Hence arose the old tollbars which were deemed such a nuisance in the middle of the nineteenth century. They would be intolerable to the swift-speeding motorist of to-day. But they had their advantages, and justly exacted tribute for the repair of the roads from those who used them.

In the old coaching days Henley and Oxford were great coaching centres. The journey from London to Oxford took two days in Anthony Wood's time, in 1665, but two years later the "flying" coaches undertook to perform the journey in one day during the summer months. The roads continued to be very bad, and the passengers by the coaches had often to dismount and walk up the hills, especially the steep ascents of Shotover. Highwaymen frequented the Chilterns, especially Gangsdown and Aston Hill, and the road between Witney and Burford was not very safe from the "gentlemen of the road." Some of the inns along the roads had not a very good reputation and murders were committed.

Matters improved with the nineteenth century. A great road led from London to Oxford and thence to Worcester. The coach passed through Brentford and Hounslow to Maidenhead, and entered the county at Henley, a town of famous coaching inns. Thence the route led by Bix Turnpike, Nettlebed, Benson, Dorchester and Littlemore to Oxford, where the four principal coaching inns were the "Angel," "King's Arms," "Roe Buck," and "Star." If you continued your journey you would pass through Wolvercote, Woodstock, Chipping

Norton, past the Four Shire Stone to Moreton-in-the-Marsh, and so through Worcestershire to its county town.

Another route from London to Oxford ran through High Wycombe, Stokenchurch, Tetsworth, Wheatley, and over Shotover Hill to Headington and Oxford. The journey had been shortened by cutting a new road from Shotover. The road would then conduct you to Witney, Burford, Northleach, and Cheltenham to Gloucester. This great road was once the glory of the county. The fame of the Great North Road, or of the Bath Road, was not greater than that of this Cotswold road. It was renowned long before the coaching age. It was traversed by the wains of the cloth merchants, and half the wealth of England passed over it to the sea. There were many cross roads along which coaches or post-chaises ran. You could journey from Bristol to Norwich, passing through Malmesbury, Faringdon, Botley Hill, and Oxford.

We are warned in *Moggs's Road Book* (1823) that the road from Oxford to Cambridge was "very indifferent," and students were advised to go round by High Wycombe and St Albans. The direct road led to Wheatley, Thame, Aylesbury, Dunstable, Royston, and Cambridge. You could also travel from Oxford to Cambridge and thence to Newmarket and Norwich by passing through Bicester, Buckingham, and Bedford.

Another road ran south from Oxford to Wallingford, and then on to Basingstoke and Chichester; another from Oxford to Derby, by Banbury and Coventry. Another connected Oxford with Peterborough, by way of Brackley and Northampton, and there were many other cross roads traversing the county.

The Great Western is the principal railway. Its main line touches the county at Goring and Culham. The Henley branch starts from Twyford in Berkshire. The northern branch leaves Didcot (Berks.) and runs nearly due north through the county from Oxford to

Canal at Oxford

Fenny Compton. A branch from Oxford goes through Yarnton and Witney to Fairford, another to Worcester by Charlbury and Chipping Norton Junction. Another runs from Chipping Norton Junction by Hook Norton to King's Sutton, and another through Wheatley and Thame to Princes Risborough, whence a branch runs to Watlington. The London and North Western Railway has a station at Oxford for a branch leading from that city to Bletchley, Bedford, and Cambridge. The Great Central Railway crosses a corner of the county and has one station at Finmere.

The opening of the Oxford and Birmingham Canal in 1790 was deemed the beginning of a new era of trade and commerce. It opened up the southern counties to the manufacture and products of Birmingham, and was especially valuable in providing a much cheaper supply of coal, which was conveyed to Oxford and the places lying north of that city, and by means of the Thames to Henley. In 1789 the Thames was connected with the Severn by the Stroudwater Canal, while the Wilts and Berks Canal from Abingdon connected Oxford with Wiltshire.

The river Thames continued to be the great means of traffic with London, but in the eighteenth century it was not the placid stream that nowadays affords delight to oarsmen. The dangers of the river in winter, in times of flood and storm, were serious obstacles. No less serious was the stranding of the barges in summer, when they lay aground for three weeks or a month. The sessions of Berkshire record many wrecks. Locks were infrequent, and the system of "flashing" was used to carry the barges over the shallows. At the shallow places stanches were placed which penned up the river, and when suddenly removed the barges were floated by the sudden rush of the water over the shallows below. The boats had to be laboriously towed up the river with the aid of a capstan on the bank.

The whole system of inland navigation was fairly

complete; millions of money had been spent in making canals and improving the course of rivers, when the invention of railways doomed them to idleness, and caused their abandonment. Possibly, now that motor traction is available, the silent waterways of England may again be used and have a prosperous future, but most of them are at present derelict and forgotten.

Mapledurham Lock

21. Administration and Divisions—Ancient and Modern.

The gradual growth and development of local government is an interesting study. Laws are made in Parliament for the whole country, but power is given to counties or parts of counties and boroughs to manage their own affairs. All this is but a development of the system of government which existed in Saxon times. Then the King was the supreme ruler, and he had a kind of central parliament, consisting of bishops, abbots, and the principal thanes or landowners, who assisted him in the governing of the country. But each county had a council called a shire-mote, which met twice a year, and consisted of the freeholders of the county. Its presiding officers were the Ealdorman, whose name survives in the modern Alderman, and the Shire-reeve, or Sheriff, who represented the King.

Besides this shire-mote there were the Hundred-courts. Each county was divided into Hundreds, each of which consisted of a hundred families of freemen. Oxfordshire has 14 Hundreds. Each Hundred had its own court, which met once a month for settling its own business, adjusting disputes, and trying prisoners. It had a regular place of meeting—some conspicuous tree, or near a ford, or a hill. Most of the Oxfordshire Hundreds are named after certain places within their area, but that of Lang-tree may have been named after some tree where the court was held.

The Saxons had yet another court. Each Hundred was composed of a number of townships or villages, and each township had its own court under the presidency of an officer called the reeve. In Norman times this became the manorial court presided over by the lord of the manor, or by his reeve or steward. These manorial courts have continued for many centuries, and their records throw much light on the history of each parish which has preserved its court-rolls.

In our own day each county has two chief officers, the Lord Lieutenant, who is appointed by the Crown and is usually a nobleman or rich landowner; and the High Sheriff, who is chosen every year on November 12th.

In 1888 County Councils were established by Act of Parliament, and the Oxfordshire County Council meets at Oxford, and consists of a chairman, 15 aldermen, and 47 councillors. They keep in repair the main roads and bridges, levy rates for this purpose and for other county requirements, manage asylums, allotments, and small holdings, and deal with education and many other important matters.

In 1894 another Act was passed which created Urban and Rural District Councils and provided Parish Councils for each parish of over 300 inhabitants. There are 16 Urban and Rural District Councils in Oxfordshire. Separated from this system of county government are the large towns, and some towns which are ancient but not large, which have their own municipal government handed down to them from a remote past. These powers have been granted by royal charters, and the towns are governed by a Mayor and Corporation. These towns are Oxford, Woodstock, Henley, Chipping Norton, and Banbury. Burford used to have its municipal corporation, but was deprived of it by the Municipal Corporation Act of 1861. Its importance was shown by the granting of sixteen royal charters to the little town.

There are also Poor Law Unions controlled by Boards

of Guardians, who manage the workhouses and administer relief to the poor. The administration of justice is at-

Banbury Town Hall

tended to by the courts of Quarter Sessions. There are also Petty Sessional Divisions, each having magistrates or justices of the peace to try cases and punish offenders.

If the crime is a grave one the prisoner is sent to be tried by the Judge of the Assize Courts.

There is a peculiar form of government at Oxford. As above stated there are the Mayor and Corporation and Borough Magistrates who try prisoners; but there is also the Vice-Chancellor's court, connected with the University. This court has great power, and exercises control over all the members of the University, who can claim to be tried in the Vice-Chancellor's Court and not by the authorities of the city. It also exercises control over the tradesmen of the city, in connection with their dealings with members of the University. The University has officers called Proctors, who have to see that the undergraduates observe the rules, and can levy fines on those who do not wear their academic dress in the streets or transgress other regulations.

There are 303 civil parishes in Oxfordshire, but for ecclesiastical purposes the county is divided into 239 parishes. Oxfordshire with Berkshire and Buckinghamshire constitute the Oxford diocese. The cathedral is at Oxford, and the Bishop resides at Cuddesdon, about eight miles from the city. Each county is an archdeaconry and is supervised by an Archdeacon. Parishes are grouped together into rural deaneries, each of which is in charge of a rural dean, whose duty it is to go round to the churches in his district and see that everything is kept in proper order.

The County of Oxfordshire sends three members to Parliament and is divided into three Parliamentary Divisions. These are the Northern or Banbury Division, the

Mid or Woodstock Division, and the Southern or Henley Division. The city of Oxford now sends only one member, but formerly returned two to Parliament, and the University of Oxford sends two members to represent it in the council of the nation.

22. Roll of Honour.

The presence of Oxford University in the shire has brought to it many of the greatest men of light and learning that England has ever known. Endless is the roll of scholars and sages, poets and professors, philosophers and theologians, who owe to Oxford their successes and their renown. But it is not possible in a single chapter to do more than refer to Oxfordshire worthies who have done honour to the county apart from the university; men and women who have been born or bred in the shire, or honoured it by residing within its borders.

Richard Cœur de Lion, writer of verses, friend of Bertrand de Born, was born in 1157 at the Palace of Beaumont outside the northern wall of Oxford. He always retained an affection for the place of his birth. Richard, Earl of Cornwall, King of the Romans, built a palace at Beckley, which parish was once owned by King Alfred, and bestowed great benefits on the shire, using his great wealth for the building and decorating of churches.

Other royal persons, natives of the county, were Edward, Prince of Wales (1330–1376), known as the Black Prince, the eldest son of King Edward III, born at Woodstock, the hero of Crecy and Poictiers; and Thomas of Woodstock, the same king's youngest son, noted for his turbulent character and his tragic death at Calais during the reign of his nephew, Richard II.

Oxfordshire can boast of some poets. The Dublin manuscript of *Piers Plowman* has a marginal note to the

Manor House, Woodstock

(*Reputed birthplace of the Black Prince*)

effect that William Langland's father was a native of Shipton-under-Wychwood, and this village is believed to have been the poet's birthplace. Geoffrey Chaucer's memory is connected with Woodstock, where he was a member of the court of Edward III. He was certainly there in 1357, and his allusion in the *Parliament of*

Fowles to a park "walled with greene stone" may be considered to refer to Woodstock. The heiress of Thomas Chaucer was the owner of Ewelme and wife of William de la Pole, Duke of Suffolk, who was murdered in 1450, and in Ewelme church is the fine marble tomb of Thomas Chaucer and his wife.

Queen Elizabeth can be claimed as an Oxfordshire poetess, as she wrote some verses during her imprisonment at Woodstock, bemoaning her hard fate, and envious of the freedom of a poor milkmaid.

Michael Drayton was not an Oxfordshire man, but he sang of the beauties of the "Isis, Cotswold's heire" that "wed with Tame, old Chiltern's son." Sir William d'Avenant, poet and writer of plays, "the sweet swan of Isis," was born at a house afterwards the Crown Inn at Oxford in 1605. The ancestors of John Milton are said to have lived at Great Milton, his grandfather at Stanton St John, and we get nearer to the poet at Forest Hill, where he stayed in 1643 and wed Mary Powell, his first wife, who brought the poet little comfort. John Wilmot, Earl of Rochester (1647–1680), born at Ditchley and educated at Burford Grammar School and afterwards at Wadham, though not a worthy, was a poet and lived at Adderbury with his Countess, to whom he often wrote when he was in London.

Pope finished his fifth volume of the *Iliad* at Stanton Harcourt, wrote an epitaph in the church, and often stayed at Mapledurham, where he imagined himself in love with Martha and Theresa Blount. Shelley was at University College, Oxford, whence he was expelled. He used to wander over Shotover, and in 1815 he tracked the Thames to its source and wrote some of his sweetest lines upon it. Wordsworth's sonnets on Oxford, written in 1820, are well known, and also his pleasant description of "a Parsonage in Oxfordshire," which is Souldern. Tennyson was married at Shiplake, where lived a clergyman who added a new word to the English language, James Granger, who wrote the *Biographical History of England*, and introduced the practice of extra-illustrating books, called "Grangerising," thereby destroying many books from which he removed plates and pictures. William Morris, poet, artist, reformer, lived in the quiet manor house at Kelmscott, mused on the fields of Isis, and laboured, sending forth into the world his printed *Odysseys* and *Aeneids*, dyed wools and wall-papers, Sagas and social pamphlets, in strange confusion. With him lived Dante Gabriel Rossetti, painter and poet, who often wrote in his poems about the natural beauties of Kelmscott. He sang of the low-lying meadows—

> "When the drained floodlands flaunt their marigolds."

He praised the beauties of the arrow-head rush with its "lovely staff of blossom just like a little sceptre"; and when the snow lay deep upon the ground, and

> "The current shudders to its ice-bound sedge;
> Nipped in their bath, the stalk reeds one by one
> Flash each its clinging diamond in the sun."

William Morris gives a description of Kelmscott in his *News from Nowhere*, and there devised his *Earthly Paradise*.

Of learned divines Oxfordshire has many. Leofric, first bishop of Exeter, was born at Bampton, which still retains its connection with the see. The bishops of Oxford have long resided at Cuddesdon, and on the walls of the Palace are the portraits of the many learned men who have held the see. It is difficult to select names of special distinction from this list of famous men, beginning with Robert King in 1545 to the late bishop, Francis Paget. Mention should be made of the heroic Bishop Skinner, who during the time of persecution under the Commonwealth lived in the rectory at Launton, and at great personal risk ordained 300 or 400 clergymen during that period. Archbishop Juxon, who attended Charles I on the scaffold, was rector of Somerton, and was buried in St John's College Chapel. Bishop Secker of Oxford, afterwards Archbishop of Canterbury, baptised, crowned, and married George III. Bishop Wilberforce's fame is so well known that it need not be recalled. Archbishop Laud's connection with St John's College has already been mentioned. Dr Heylin, theologian and historian, chaplain of Charles I, was born at Burford. The Puritan divine, John Owen, was born at Stadhampton, of which parish his father was minister. He had to leave Oxford on account of his opposition to Laud's regulations. He preached before Parliament on the day after the execution of Charles I, and became Cromwell's chaplain, and subsequently Dean of Christ Church, Oxford, and Vice-Chancellor of the University. Dr South, the witty preacher of the Restoration period, was rector of Islip, and left behind him some memorials in a rebuilt chancel and rectory. Islip was famous for its learned rectors, who were also Deans of Westminster. Amongst them may be mentioned William Vincent, philologist, died 1815, John Ireland (1761–1842), founder of the Ireland scholarships at Oxford, and Dr William Buckland (1784–1856), the great geologist, and father of Frank Buckland, the naturalist. John Henry Newman, afterwards Cardinal, was vicar of Littlemore and also of St Mary's, Oxford, just before his secession to the Church of Rome. The church at Littlemore was built by him in 1836.

Of statesmen and warriors Oxfordshire has had no lack. William Marshall, Earl of Pembroke, the guardian of Henry III, died at Caversham. At Ewelme lived William de la Pole, Duke of Suffolk, who died a violent death at sea in 1450, and afterwards Edmund, Earl of Lincoln. Henry VIII visited him at Ewelme but afterwards executed him. Lucius Cary, Lord Falkland, one of the noblest men of the Civil War period, was born at Burford in 1610, killed at the battle of Newbury, and buried at Great Tew, where his home was. Sir Henry Lee, K.G., lived at Ditchley, where his descendant, Lord Dillon, now resides. He performed good service at the siege of Edinburgh in 1574 and was made Ranger of Woodstock Park. Dudley Carleton, Viscount Dorchester (1574–1632), second son of Anthony Carleton, was born at Baldwin Brightwell, and was ambassador in Venice, Holland, and France, and Secretary of State. He was a great patron of painters, and to him Rubens dedicated the engraving of "The Descent from the Cross." The family of Saye and Sele lived at Broughton

Castle, and has produced many warriors, statesmen, and illustrious men. The Knollys family lived at Grey's Court and Caversham. Sir Francis was treasurer to the court of Queen Elizabeth. His son, Sir William, entertained Queen Elizabeth and James I at Caversham, and played a distinguished part in the public affairs of the period. The Harcourts of Nuneham and Stanton Harcourt, which they held for over seven centuries, have produced many statesmen and soldiers. Sir Simon Harcourt was Lord Keeper of the Great Seal in 1702, was ennobled by Queen Anne and appointed Lord Chancellor. Stanton was obtained by the marriage of Robert de Harcourt with Millicent de Camville, kinswoman of the Queen of Henry I. Edmund Vernon Harcourt, Archbishop of York, was the inheritor of the estate of Nuneham Courtenay in 1830. John Hampden, the patriot, wooed and won Squire Symeon's daughter of Pyrton Manor. The Cottrells and Dormers have played a distinguished part in the nation's history. The founder of the Dormer family was Geoffrey, a successful merchant of the Staple at Calais who lived at Thame, where in the church is his monument. He had a large family of twenty-five children. Westminster Abbey contains a memorial of Clement Cottrell, who fought gallantly in the fight with De Ruyter off Southwold Bay. His father was page to George Villiers, Duke of Buckingham, steward of Elizabeth, Queen of Bohemia, and after the triumph of Cromwell took charge of the younger son of the King.

Lord North, second Earl of Guildford (1732–1792),

Prime Minister of George III, lived at Wroxton Abbey. He held office during the American War, and was always treated by his sovereign as a personal friend. A brass memorial at Whitchurch records the memory of a bold man, Sir Thomas Walysch, who was valet-trayer or food-taster to the Lancastrian monarchs, Henry IV, V, and VI. In the days of conspiracies and poisons his office required some courage.

Regicides should not appear on a roll of honour, but Sir James Harrington of Merton who took a prominent part in the Civil War and was one of the king's judges, must be mentioned as a notability. He resided for a long period at the manor house of Merton, and his grandson, Sir James Harrington of Merton and Caversfield, was a vehement Jacobite, is said to have entertained Charles Stuart after Culloden, and after 1745 lived in France at the court of the Pretender. Another regicide, Adrian Scrope, owned Wormsley, and at the Restoration was executed. Sir John Borlase, who owned Stratton Audley, an Irish Judge, helped to suppress the Irish rebellion of 1641. His grandson, Admiral John Borlase Warren, fought with distinction in the American and French naval war in the eighteenth century, and lies at Stratton.

Warren Hastings was born at Churchill in 1732 and was brought up as a charity boy at the parish school.

Robert Dudley, Earl of Leicester, one of Queen Elizabeth's earliest favourites, and the only one who succeeded in retaining her favour to the close of his career, died at Cornbury House, and was supposed to have been poisoned by his second wife, Lettice Knollys. If the story be true that he caused the death of his first wife Amy Robsart, at Cumnor, as recorded in Sir Walter Scott's *Kenilworth*, he deserved his fate. At Cornbury lived Edward Hyde, first Earl of Clarendon (1609–74), Lord Chancellor of England, and the historian of the great Civil War, and his

Warren Hastings

son, Lord Cornbury, took his title from that place. He was a member of the Long Parliament, and took an active part in trying to reform abuses, but adhered to the royal cause when the Civil War broke out. His daughter

Anne married the Duke of York, afterwards James II, and thus he became the grandfather of two Queens of England, Mary and Anne. This marriage created for him many enemies, and he died in exile at Rouen. At Godstow lived Sir John Walter, a famous lawyer, who was Attorney-General to Prince Charles in 1613, and Chief Baron of the Exchequer. In the matter of the dispute between the King and the Commons he lost the favour of his sovereign by his uprightness and died in 1631. His monument is in the church at Wolvercote. Sir George Coke, of Studley, was a noted judge in the Court or King's Bench in Charles I's reign, and was bold enough to pronounce against the legality of the King's attempt to extort ship-money. He bought Waterstock, and died there in 1641.

At Burford Priory at the close of his life lived William Lenthall, Speaker of the House of Commons, who presided over the Long Parliament. He was a native of the county, having been born at Henley-on-Thames in 1591, was educated at Oxford at St Alban's Hall, and represented an Oxfordshire constituency, Woodstock. He bought Burford Priory just before the Civil War, and his family held it nearly 200 years. As Speaker he lived in dangerous times and occupied the chair when Charles I entered the House, and demanded the surrender of the five members. He welcomed the Restoration of the monarchy, but retired to his beautiful home at Burford and died two years later. In the same house lived Sir Lawrence Tanfield, Chief Baron of the Exchequer in the reign of James I.

Shirburn Castle was bought by the celebrated Thomas Parker, Lord Chief Justice and Lord Chancellor in 1718, who was created Earl of Macclesfield, and afterwards impeached by the Commons; it has since been held by other distinguished members of this family.

William Lenthall

Thame has produced a just judge whose praise Steele chants in the *Tatler*, Lord Chief Justice Holt, born in that town in 1642.

Amongst other successful lawyers whom Oxfordshire has produced was Roundell Palmer, first Earl of Selborne

(1812–95), who was the second son of the rector of Mixbury, where he was born. After a brilliant career at Oxford he became Solicitor-General in Lord Palmerston's ministry in 1861 and Lord Chancellor of England. He published *The Book of Praise*, and loved hymns and verses quite as much as law books. Another great lawyer, Sir John Bankes (1589–1644), though not a native of the shire, lived at Oxford during the Civil War period when Charles I held his court there, and died there and rests in the cathedral. He was Attorney-General in 1634 and earned the reputation of having exceeded "Bacon in eloquence, Chancellor Ellesmere in judgment, and William Noy in law." Oxfordshire has been famous for its successful lawyers.

Besides the warriors we have already mentioned we may record the name of Earl Cadogan, one of Marlborough's most trusted generals. "Cadogan's Horse" became famous on many battlefields, and he won renown at Blenheim, Ramilies, Oudenarde, and Malplaquet. He crushed the Stuart rising in 1715, and on the death of the Duke he became commander-in-chief of the armies of England. He resided for some years at Caversham Park. The famous John Churchill, first Duke of Marlborough, himself, of course, resided at Blenheim, the gift of a grateful nation, where he was buried, and there lived also his no less famous Duchess, Sarah Jennings, the favourite friend and Keeper of the Privy Purse of Queen Anne. Her violent temper used to try the Duke, and even her royal mistress found her unbearable at times, and she lost her favour through the intrigues of Lady

Masham. The last soldier we will mention is Sir John Cope, the ill-fated commander at Prestonpans in Scotland, a battle fought in 1745 against Charles Edward the "Young Pretender," who defeated the royal troops and invaded England. Sir John, or as he was usually called,

Wadham College
(*Where the Royal Society had its origin*)

Johnnie Cope, lived at Bruern Abbey, which was afterwards destroyed by fire.

Roger Bacon, early scientist and philosopher (1214–92), lived and worked, and probably died at Oxford. Friar Bacon's study was the name of an old gatehouse near Folly Bridge, pulled down in 1770.

The presence of the court of Charles I at Oxford brought together many men of eminence in various professions. We like to think of the celebrated Dr Harvey (1578–1657) during the turmoil of war quietly conducting his experiments on the circulation of the blood, at Merton College, of which he was appointed Warden by Charles I. He set a hen in a corner of his room and every day broke an egg to see how it had developed. During this time of anxiety and continual fighting the Royal Society practically commenced its career at Wadham College. Dr Harvey enjoyed the confidence of the King, and during the battle of Edgehill was entrusted with the care of the royal children, the Prince of Wales and the Duke of York.

Oxfordshire has had the honour of producing several artists of distinction. Sir William Beechey, R.A. (1753–1839), who painted portraits of most of the royal persons of his day and also of the most fashionable, literary, and theatrical characters of the period, was born at Burford. Sir William Roxall, R.A. (1800–79), another celebrated portrait painter, and director of the National Gallery, was the son of an Oxfordshire exciseman. Waller came of an old Burford family and introduced the old priory into his painting of "The Empty Saddle" and into some other pictures. The wife of Sir Godfrey Kneller was the daughter of Mr Cawley, rector of Henley, where she lies buried with her parents. Valentine Green (1739–1813), associate-engraver of the Royal Academy, antiquary and author, was born near Chipping Norton, being the son of a dancing-master. He achieved fame as a mezzotint-engraver, reproducing by his art some of the best portraits of Sir Joshua Reynolds. Another mezzotint-engraver was Charles Turner (1773–1857), who was born at Woodstock. He was one of the most skilful engravers of his time, and was associated with the world-renowned land-

Sir John Soane

scape painter, Joseph Mallord William Turner, R.A., in the earlier stages of the *Liber Studiorum*. The distinguished architect, Sir John Soane, R.A. (1753–1837), was an Oxfordshire man, having been born at Whitchurch, the son of a mason named Swan. He preferred to call himself Soane, became a pupil of George Dance, the architect, and by his industry and skill raised himself to the head of his profession. His famous Museum, which he founded in Lincoln's Inn Fields, he endowed and bequeathed to the nation.

Dr White Kennet, afterwards Bishop of Peterborough, the author of the *Parochial Antiquities of Ambrosden and Burcester*, a celebrated Oxfordshire book, was vicar of Ambrosden in the seventeenth century. Dr Plot, whose words we have more than once quoted, author of the *Natural History of Oxfordshire* in 1677, though not a native, lived long at Oxford as Keeper of the Ashmolean Museum. Other local historians were the Rev. J. C. Blomfield, rector of Launton, author of the history of Bicester Deanery, and the Rev. Edward Marshall, author of the Diocesan History of Oxford.

These are some of the names of the most illustrious sons of Oxfordshire, a roll of honour of which any county might be proud.

23. THE CHIEF TOWNS AND VILLAGES OF OXFORDSHIRE.

(The figures in brackets after each name give the population in 1901, but where the figures of the 1911 census are available they are given, preceded by an asterisk. The figures at the end of each section refer to the pages in the text.)

Adderbury (2025), three and a half miles south of Banbury, is a picturesque town with a fine church remarkable for its spire. There is an old rhyme:—

> "Bloxham for length,
> Adderbury for strength,
> And King's Sutton for beauty."

The principal feature of the church is its beautiful Decorated work and the Perpendicular chancel, erected by the famous inventor of this style, and foremost architect of his age, William of Wykeham. The witty John Wilmot, Earl of Rochester (1647-80), and the second Duke of Argyle lived here in an old house on the town green, where Pope stayed. (pp. 40, 68, 116, 118, 175.)

Alchester, the remains of a Roman town two miles south of Bicester. (pp. 102, 162.)

Asthall (355) is a beautiful village three miles south-east of Burford, near the Roman Akeman Street, with an interesting church and an Elizabethan manor house. The church retains some Norman features, and is late Norman in construction with additions and decorations made in later periods. (pp. 109, 142, 162.)

Bampton (2029), an old-world market town with a noble church. At one time a castle stood here, built by Aymer de Valence, Earl of Pembroke, in 1315. A skirmish in the Civil War took place here. Many old customs have lingered on in this

Globe Room, Reindeer Inn, Banbury

obscure place. The church is particularly fine, notable for its massive Norman work and Early English additions. The spire belongs to the latter period, and resembles that of the cathedral. The Early English sedilia, Decorated reredos, Perpendicular Easter sepulchre and misereres are interesting features. (pp. 88, 108, 116, 128, 130, 131, 177.)

Banbury (*13,463) is a large and prosperous town, celebrated for its cakes, its cross, and its Puritan zeal. Agricultural implements are made here. It has a large market. A castle once stood here, which played an important part in the Civil War and was destroyed at its close. A very noble church formerly existed, but it was ruthlessly destroyed at the beginning of the nineteenth century, and a hideous structure erected in its place. There are several picturesque half-timbered buildings, and some noted inns, including the "Old Reindeer" which contains a remarkable room

Benson Weir

called the Globe Room. (pp. 15, 23, 33, 40, 67, 68, 77, 80, 84, 85, 90, 93, 128, 129, 160, 161, 165, 170–2.)

Beckley (573) is an ancient and picturesque village five miles north-east of Oxford, once held by King Alfred. Roman remains have been discovered here. It was an important place in ancient times and the manor has been held by Robert D'Oilly, Richard King of the Romans, Piers Gaveston, Hugh le Despenser, and Lord Williams, one of the church spoilers at the Reformation. The church has some interesting features, including some old glass, a holy-water stoup and an hour-glass stand. The choir is Decorated c. 1330. (pp. 10, 90, 102, 173.)

Bensington or **Benson** (960), on the Thames a mile north of Wallingford, was a busy town in the coaching age with large inns and much traffic. It is now only a village. A British town stood here which was taken by the Saxons in 571, and Offa King of Mercia defeated the West Saxons here in 777 A.D. The old late Norman church has been much modernised and restored without much respect to its old features. (pp. 86, 164.)

Bicester (*3385) is a small market town with a notable church. It had at one time a priory, founded by Gilbert Bassett in 1183, of which there are some scanty remains. Its name was formerly Berenceastre, and is probably connected with St Birinus, the apostle of Wessex. The church has a Saxon doorway, the remains of an earlier structure, and much Norman work, though the aisles were added in the thirteenth and fourteenth centuries. (pp. 10, 16, 34, 42, 53, 67, 85, 108, 122, 162, 165.)

Bletchingdon (549), six miles north of Oxford, is a small village which owes its interest to the Park, the seat of Lord Valentia. The modern house replaces one that was held for the King in the Civil War and captured by Cromwell.

Bloxham (1509), three and a half miles south-south-west of Banbury, is a pleasant village that calls itself a town, and is noted for its school and its magnificent church. The rhyme relating to "Bloxham for length" has already been quoted under Adderbury. This church is one of the most beautiful in the county. There is some Norman work, and also Early English, but its chief feature is the splendid Decorated work of the north side of the church, and of the spire and tower. The manor was acquired by the Fiennes family in the reign of Henry VIII and is still owned by Lord Saye and Sele. (p. 111.)

Broughton (525), two and a half miles south-west of Banbury, is a small village with a Decorated church, and is rendered famous by the magnificent castle, the seat of Lord Saye and Sele. It was first built by the De Broughtons at the beginning of the fourteenth century. William of Wykeham bought the manor and presented it to his great-nephew Sir Thomas Wykeham, whose heiress brought it by marriage to William, Lord Saye and Sele. Thomas Wykeham in 1467 erected part of the gatehouse and other portions, and the Fiennes family added the Elizabethan buildings. There is a chapel in the house of the Decorated period. In the council chamber William, Lord Saye and Sele, and his friends are said to have devised plans for the resistance to the King which led to the Civil War. A long room in the attics, wherein the soldiers of Lord Saye and Sele slept before the battle of Edgehill, is called the Barracks. (pp. 93, 116, 118, 134–6, 141, 142, 178.)

Burford (1323) is one of the most interesting towns in the county. It had once a mayor and corporation, but has now lost its municipal rights. It was formerly famous for its cloth, wool, stone, malt, and saddles. It stood on the borders of Wessex and Mercia and was the scene of many fights. Near it was the Forest of Wychwood, the hunting ground of Norman kings, and it was the first town in England to receive the privilege of a Merchant Guild. Earl Warwick, the King-maker, owned the town, built the porch of the church, and founded the almshouses, though they were really given by Henry Bishop, a native of the town. A priory stood here which at the dissolution of the monasteries was given to Edmund Harman, King Henry's barber-surgeon. Afterwards it passed to Sir Lawrence Tanfield, then to Lucius Cary, Lord Falkland, and then to William Lenthall, Speaker of the Long Parliament. Burford was visited by Queen Elizabeth, James I, Charles I, Charles II, and William III. The church is a magnificent structure, and near it are the almshouses and Grammar School. Of this church, Street, the architect, says that there is not one in the whole diocese of Oxford which exceeds it in beauty and architectural interest. It was originally cruciform, with a central tower; but so many aisles and chapels have been added that its plan is irregular. Little Norman work is visible, and the architecture is mainly of the Early English and Perpendicular style. It is replete with details of architecture, tombs, monuments, and many objects of unusual interest. The old inns, the "Bear" and the "George," the ancient Tolsey or town-hall, and many old gabled houses, are all worthy of notice. The town

Hampden's Obelisk, Chalgrove

was the scene of the rebellion of the Levellers against Cromwell. It was suppressed here, some of the men being shot in the churchyard. (pp. 8, 17, 22, 29, 39, 40, 42, 54, 68, 70, 77, 80, 86, 93, 109, 121, 125, 144, 164, 170, 175, 177, 178, 182, 186.)

Caversham (*9858) is now a suburb of Reading, and has largely increased its population. A fine medieval bridge, upon which was a chapel, crossed the Thames, but it has been replaced by a hideous iron one. A fight took place on the bridge in the Civil War. A priory stood here, a cell of Notley Abbey, the buildings of which form part of the rectory, and St Anne's Well was much frequented. The Knollys family lived at Caversham Park, where Queen Elizabeth and James I visited, and Charles I was retained a prisoner. The old house was destroyed by fire and a new one erected. (pp. 12, 48, 67, 80, 84, 85, 96, 178, 179, 184.)

Chalgrove (379), three miles west-north-west of Watlington, is a small village near which the battle of Chalgrove was fought in 1643, wherein John Hampden was mortally wounded. An obelisk marks the spot. (pp. 34, 93, 116.)

Charlbury (2711) is a quiet little town on the banks of the Evenlode which has had several lords of the manor, the Bishops of Lincoln, Eynsham Abbey, Sir Thomas White, and St John's College, Oxford. Near it is Lee Place, a Jacobean mansion formerly owned by the Lees of Ditchley, and Cornbury House, adjacent to Wychwood Park. The church is Norman and Early English. (pp. 68, 94, 115, 166.)

Chinnor (1002) is a village situated in the midst of the beautiful scenery of the Chilterns three and a half miles west of Princes Risborough, near the Buckinghamshire border and within sight of the turf-cut monuments of Whiteleaf and Bledlow Crosses. It has several old houses and a late Decorated church, which has some remarkable brasses of the fourteenth and fifteenth centuries. (pp. 80, 116, 162.)

Chipping Norton (*3972) is an old market-town, as its name implies, Chipping being derived from the Saxon word "to buy." A castle was erected here in Stephen's reign. King John granted a market to the town and it returned members to Parliament in the time of Edward I. James I granted a charter to the town. It has a noble church, principally of Decorated and Perpendicular work, one of the finest in the county. The town-hall is modern, but the old guild-hall still exists. There are almshouses erected in 1640. The principal trades of the town are tweed-making and brewing. (pp. 8, 31, 33, 40, 67, 77, 118, 128, 130, 163, 164, 166, 170, 186.)

Coggs (790) is an interesting village near Witney, remarkable for its group of old buildings, church, vicarage, barn and manor house. (pp. 121, 128, 138.)

Cowley (9258) owes its large population to its nearness to Oxford, of which part of the parish is a suburb. The villages of Temple Cowley and Church Cowley are two and a half miles from the city. The church is curious, as the tower is not as high as the roof of the nave, but it has some good Norman and Early English work. The beautiful chapel of the Hospital of St Bartholomew, the property of Oriel College, once a leper's hospital, is in Cowley. (p. 108.)

Cropredy (436) is a small village three and a half miles north of Banbury where a battle was fought in the Civil War. It has a good church, principally of the Decorated period, which contains some armour of soldiers slain in the battle. (pp. 93, 116.)

Cuddesdon (1332) contains the palace of the Bishop of Oxford, originally built in 1635, burnt in the Civil War, and rebuilt in 1679. The manor belonged to the Abbey of Abingdon. The church is cruciform, of Norman date, with thirteenth century aisles and Perpendicular chancel. (pp. 44, 112, 172, 177.)

Deddington (1490) is a small market-town and once possessed a castle. Here Piers Gaveston was captured by the Earl of Warwick. Castle Farm, formerly a sixteenth century rectorial house, is an interesting building where Charles I slept after the battle of Cropredy. Sir Thomas Pope, the founder of Trinity College, Oxford, and Chief Justice Scroggs, the hanging judge, were born here. The church is large and principally of the Decorated period. Charles I made cannon out of its bells. (pp. 40, 128, 130, 131, 132.)

Dorchester (944) is a small village nine miles south-south-east of Oxford, once the seat of a far-extending diocese reaching from the Thames to the Humber. Cynegils granted it to St Birinus, the apostle of Wessex. Six bishops ruled here from 634 to 706.

Dorchester

In 869 it was a Mercian see and five bishops ruled in succession until Lindsey and Leicester were joined with it, when 10 bishops held the see until the Norman bishop Remigius transferred the seat of the bishopric to Lincoln. A priory existed here, founded in 1140, of which the present church was the minster. It is an extremely interesting building principally of the Decorated style.

It has a leaden Norman font with carved figures, and a Jesse window, showing the descent of Our Lord from the stem of Jesse. (pp. 65, 87, 102, 112, 115, 162, 164.)

Enstone (925), which takes its name from the dolmen called the Hoar Stone, lies four miles east-south-east of Chipping Norton. It belonged to the Abbey of Winchcombe, Gloucestershire. The fine tithe barn was built by Walter de Wyniforton, abbot in 1382.

Ewelme

The church has a curious dedication, to Kenelm, son of Kenulph, King of Mercia. The village was a great coaching centre with six inns. (pp. 99, 112.)

Ewelme (494) is a very picturesque village between Watlington and Wallingford with a grand group of old buildings, consisting of church, almshouses, and school, erected by William de la Pole, Duke of Suffolk, in 1436. His tragic story has been told already. The church is entirely Perpendicular and has some interesting monuments. The hospital or almshouse and the school are built of brick, and are very beautiful. The Manor Place was of timber and brick, but has been pulled down. A brook runs through the village and supplies many water-cress beds. This water-cress is famous and is sent to London and Lancashire. (pp. 118, 122, 175, 178.)

Goring

Fritwell (454), five miles north-west of Bicester, has a church dedicated to St Olaf, and therefore was doubtless connected with the Danes. There is an old manor house here which has been connected with some unhappy tragedies. (pp. 111, 142.)

Godstow, three and a half miles north-west of Oxford, has the ruins of the nunnery where dwelt Fair Rosamond. Converted into a dwelling-house at the Dissolution, it was fortified for the King in the Civil War, besieged and burnt in 1646. An old bridge spans the river and the "Trout Inn" is famous. (pp. 53, 88, 122, 182.)

Goring (1419) is noted for the beauty of its river scenery, which has attracted many new residents and caused the erection of many modern houses. There was a nunnery here, founded in the reign of Henry II. The church is dedicated to St Thomas a Becket, and was used by the nuns as a chapel. (pp. 27, 80, 85, 162, 166.)

Hanwell (176) lies two and a half miles north-west of Banbury, and has a farm which is all that remains of the Castle of Hanwell, the house of the Copes, erected by Anthony Cope, cofferer to Henry VII. The church was built in the thirteenth century and has some Decorated work.

Haseley, Great (551), five miles south-west of Thame, has been held by many distinguished families, including Milo Crispin, the Bassetts, Roger Bigod, Thomas de Brotherton, the Bohuns, the Lenthalls. John Leland was rector here, and also Christopher Wren, the father of the architect. The thirteenth century church has some good brasses. (pp. 115, 116.)

Headington (3696) with its hill is well known to Oxford men as a favourite walk. The road is celebrated as that traversed by the royalist garrison of Oxford when they marched out of the city and delivered it to Cromwell's troops. A legend tells of a student being attacked here by a wild boar and saving his life by cramming his book of Aristotle down the animal's throat, saying *Graecum est*. This is said to have been the origin of the custom of bringing in the Boar's Head on Christmas Day at Queen's College. At Headington there stood the royal hunting lodge of Saxon kings. The church is mainly Norman with Perpendicular tower, and a fine sixteenth century cross stands in the churchyard. The Headington quarries have supplied much stone for the building of the Oxford colleges. (pp. 8, 22, 43, 84, 85, 88, 165.)

Henley (*6456) is a fair town on the Thames, noted for its Regatta and for the beauty of its surrounding neighbourhood. Richard Earl of Cornwall, King of the Romans, had a palace here in the reign of Henry III. The manor passed to Edward I who granted it to Piers Gaveston, and after his execution it passed to Sir John de Moleyns, in whose family it remained for

Church, and Red Lion Inn, Henley

some time, and then by marriage passed to the Hungerfords, and then to Lord Hastings in the reign of Henry VII. It has since been held by several notable men. The town suffered much during the Civil War, and Sir Bulstrode Whitelock, who lived at Fawley Court, complained bitterly of the damage done by the soldiers. Charters were granted to the town by Henry VIII,

Elizabeth, and George I, and it is governed by a mayor and corporation. The town-hall was built in 1795 on the north of the market-place. A fine bridge spans the river, erected in 1786 in place of a very early structure, on which was the chapel of St Anne. The church is mainly Decorated and Perpendicular with a fine tower. The town has several old inns, as it was a great centre in the coaching days. The "Red Lion" Inn was in

Iffley Church

existence in the time of Charles I, who stayed there with Prince Rupert. It has had many other royal and illustrious guests. The "Catherine Wheel," "White Hart," "Bull," "Bear" and "Broad Gates" are other ancient hostelries. The present trades of the town are malting, brewing, and the making of boats and paper-bags. There is also an iron foundry. (pp. 8, 16, 23, 27, 28, 47, 72, 77, 80, 101, 132, 136, 163, 164, 167, 170, 173, 182, 186.)

Hook Norton (1386), four and a half miles north-east of Chipping Norton, is a growing village owing to the ironstone industry. The Danes were massacred here by Edward the Elder. (pp. 18, 35, 40, 68, 88, 112, 160, 166.)

Iffley (2358), a beautiful village on the Thames near Oxford, famous for its scenery and its fine Norman church. (pp. 26, 109.)

Islip (549), at the junction of the Ray and Cherwell five miles north of Oxford, had a royal Saxon palace in which Edward the Confessor was born, who gave the manor to Westminster. There was much fighting here in the Civil War. (pp. 88, 177, 178.)

Kelmscott, a small village on the Thames where the river enters the county, the residence of William Morris and Rossetti. The name is a shortened form of Kenelm's Cot, the abode of St Kenelm, from whom Kempsford in Gloucestershire also derives its name. Minster Lovell church is dedicated to the same saint, who was the son of Offa, King of Mercia, and when a child was murdered by his tutor at the instigation of his sister. (pp. 12, 16, 112, 176.)

Kidlington (1380) is a large village, four miles north of Oxford, remarkable for its noble church with a lofty spire. The tower was erected in the thirteenth century and crowned with a spire in the fifteenth. The main style of the building is Early English, but additions and alterations were made later in the Decorated and Perpendicular period. It will be noticed that the chancel leans to one side. This is traditionally said to represent the leaning of the head of the Saviour on the Cross. (pp. 33, 34, 72, 114, 116.)

Kirtlington (594), seven miles north of Oxford, was a village of importance in early times on account of its position at the junction of Akeman Street and the Portway. An important synod was held here in 977. (pp. 88, 161, 162.)

Langley, five miles north-west of Witney, had a royal palace or hunting lodge, said to have been built by King John, and visited by many kings and queens until the reign of the Stuarts.

Mapledurham (519), a picturesque village on the Thames with a famous manor house, the home of the Blounts, which was besieged during the Civil War. The church has a Norman font and is mainly Perpendicular. The chapel of the Blounts is walled off from the rest of the church. (pp. 27, 136, 142, 175.)

Minster Lovell (459), two and a half miles north-west of Witney, has the ruins of the manor house of the Lovells. Reference has been made to the sad fate of the last of the family. There was a priory here; hence its name Minster, which is always connected with a monastic house. Near the village are the Charterville Allotments, founded by Fergus O'Connor, the Chartist leader in 1847. The scheme failed, but is now successful, and some labourers support themselves by growing strawberries and potatoes. (pp. 19, 30, 90, 118, 121, 140.)

Northleigh (717), three miles north-east of Witney, has near it the remains of an extensive Roman villa. (p. 102.)

Otmoor is a desolate region on the eastern border of the shire, formerly consisting of bog and moorland, which fed great flocks of geese. It was drained and enclosed in 1830, when wild rioting ensued, the Yeomanry were called out and violent scenes took place, and for four years disturbances were frequent. (pp. 21, 34, 162.)

Oxford (*53,049, this does not include members of the University, the last census being taken in Vacation time). Oxford, the county town, lies between the river Cherwell and the Thames or Isis, and a small part of it is in Berkshire. It is 63 miles from London by rail, and lies centrally in the county at its most constricted part. The city is renowned for its charm; its High

Street is held to be one of the most striking thoroughfares of Europe; and everywhere are seen churches, colleges, quadrangles and palatial buildings which are remarkable for their architectural beauty. In early times its geographical position at the head of a great river rendered it important, though it lay far from any of

High Street, Oxford

the Roman roads, and in Roman times cannot have been in existence. The foundation of the nunnery of St Frideswide about 730 was the earliest event of importance in its history. It grew into prominence two centuries later when there was peace between Wessex and Mercia, its prosperity being increased by its ford across the Thames and the traffic along the road leading from

Pulpit in the Quadrangle, Magdalen College

Berkshire to Oxfordshire. Edward the Elder and his heroic sister Ethelfleda probably raised the great mound on which the Norman castle was afterwards constructed and fortified it against the Danes. There was a large colony of Danes here, and when in 1002 King Ethelred ordered a massacre of the Danes on St Brice's Day, many took refuge in St Frideswide's convent, which was set on fire, and great numbers perished. In 1013 they revenged themselves and Sweyn, their king, captured Oxford. Many important gemots, or councils, were held here, and here most of the Saxon kings sojourned, and some died. When the Normans came, William the Conqueror attacked the town in 1067, and Robert D'Oilly, to keep the country in awe, built the strong stone fortress that still stands near the railway station. This castle became the scene of many historical events; State councils were held there, and during the wars of Stephen's reign it was besieged, and the Empress Maud escaped from it by night dressed in white when snow was on the ground. The royal palace of Beaumont, which stood where Beaumont Street now is also added importance to the city, where Kings Richard and John were born. The history of Oxford is closely interwoven with the history of England, especially from the time of Henry III. During his reign the walls of the city were built in stone, the *Provisions of Oxford* were passed and the University was founded, though Schools and Halls for students had previously existed. A large migration of scholars from the University of Paris swelled the number of the students, who lived in halls and private houses in the town. They were often very turbulent, frequent riots took place and conflicts between the scholars and the townspeople, the most serious being that which occurred on St Scholastica's day in 1354, when forty students and twenty-three townsmen were slain. Colleges were gradually built for the reception of scholars, Merton, transferred to Oxford in 1274, being probably the earliest collegiate foundation. Learning thrived, and many distinguished foreigners were attracted to Oxford by its fame. An interesting feature of

Interior: Christ Church Cathedral

Magdalen College (founded in 1458) is the open pulpit in the corner of the small quadrangle. One of the earliest printing-presses in England was set up in the town; and its library was celebrated for its wealth of books and manuscripts. The Dissolution of monasteries in the reign of Henry VIII interfered much with the progress of learning. Monastic houses throughout England were accustomed to send poor scholars to Oxford for their education. This stream of students was at once stopped. Osney Abbey, St Frideswide's Monastery, and several houses of the Friars were suppressed, and the valuable contents of the University Library were burned, on the ground that the books were of a "superstitious" nature. However, the King completed the great design of Cardinal Wolsey in the foundation of Christ Church, and constituted Oxford a Cathedral City. Soon followed the Marian persecution and the burning of the martyrs Cranmer, Ridley, and Latimer. Oxford took a prominent part in the Civil War in the seventeenth century, when the city was the headquarters of Charles I. It had great affection for the Stuarts, and Jacobitism lingered long there. Oxford has progressed much in recent years, and has opened its gates wide to welcome all classes of scholars, providing an education which, while preserving the heritage of the past, is suitable to the needs of modern life. There are no manufactures of importance, nor is there much trade except such as is connected with the supplying of the University and other inhabitants, the place of late having shown increasing popularity as a residential neighbourhood. The University Press, however, employs a large number of men. There are 21 Colleges and one Hall with a roll of about 3000 undergraduates, and about 500 resident graduates. (pp. 2, 8, 10, 12, 14, 22, 23, 25, 26, 62, 63, 65, 72, 73, 75, 76, 77, 80, 84, 85, 88, 90, 92, 93, 102, 106, 108–14, 116–9, 122–9, 138, 144, 145–59, 161, 164–7, 170, 172, 173, 175–8, 182, 184–6, 188.)

Rollright is famous for its remarkable Stone Circle which has already been described. (pp. 98–101, 109, 111, 160, 161.)

Rotherfield Greys (3337, including a portion of Henley town parish) derives its name from the family of Greys who owned the castle. The adjoining village of Rotherfield Peppard took its name from the ancient Pypard family. The castle and court have already been described (p. 132). The property is now owned by the Stapleton family. The church is mainly Early English work and has been much restored. (pp. 65, 132.)

Rycote, two miles west-south-west of Thame, once had a notable mansion owned by the Norris family. Princess Elizabeth was detained here under the guard of Lord Williams during the rule of her sister Mary. Charles I stayed at the house, which has now been destroyed. (pp. 72, 92, 120.)

Shiplake (870) is a beautiful village on the Thames three miles south of Henley. The manor was held by the Blunden and Plowden families. James Granger whose name gave a new word to the English language—that of "Grangerising" or extra-illustrating books—was vicar here, and Tennyson was married in the church. In the church there is some fine glass, brought from the Abbey of St Bertin at St Omer, when it was destroyed by the French revolutionists. (pp. 48, 80, 176.)

Shipton-under-Wychwood (2686) is noted for its fine church, which has a notable tower and spire similar in style to those of Oxford Cathedral. It is mainly Early English with some sixteenth century additions. There are two ancient inns, the "Red House" and the "Crown." Shipton Court is an Elizabethan house, formerly the home of the Reades. A house called the Prebendary recalls the fact that the town was formerly attached to the Cathedral of Salisbury. (pp. 31, 112, 122, 174.)

Somerton (265) is a village by the Cherwell seven miles north-west of Bicester and once had a castle, erected in Stephen's reign. (pp. 128, 177.)

Standlake has a fine cruciform church and near it the remains of a British village. (pp. 31, 101.)

Stanton Harcourt (491), four and a half miles south-east of Witney, is named after the family who held it for many centuries. Little of the old manor house remains, but there is the gatehouse and Pope's tower, and a very interesting church. It is an epitome of Gothic architecture, showing Norman work in the nave, Early English in the chancel, a Decorated roof, Perpendicular windows, and a Tudor chapel. The stocks remain in the village. To the south are the prehistoric stones known as the Devil's Quoits. (pp. 65, 99, 108, 114, 140, 175, 179.)

Stanton St John (452), four miles north-east of Oxford, takes its name from the family of St John. The church was mainly built at the beginning of the fourteenth century, though it has a Norman chancel arch and Perpendicular west tower. (pp. 65, 116, 175.)

Stonor Park, some four miles from Henley, has always been owned by a family of that name. Sir Thomas Stonor commanded the left wing at Agincourt and was created Baron Camoys. The present house is Tudor, and near it is a fourteenth century chapel wherein Roman Catholic services have always been held. There are priests' hiding-places cunningly devised in the house, and there Edmund Campion the Jesuit lay safe and established a secret printing-press. (pp. 16, 93, 144.)

Tew, Great (334), a singularly beautiful village five and a half miles north-east of Chipping Norton, full of the memories of the old house where Lucius Cary, Lord Falkland, lived and his loving wife Lettice. The church is ancient, with a Norman door and early fourteenth century work, during which period the main part was built in the Decorated style. (pp. 136, 178.)

Thame (*2957) is a market-town of great antiquity and belonged to the bishops of Lincoln. It is indebted to one of

Witney Butter Cross

their bishops, Robert Grosseteste, for its magnificent church and Prebendal chapel. The church retains much of Bishop Grosseteste's work, erected about 1240, but many alterations have been made since. It contains many monuments of interesting persons. Lord Williams, who grew rich out of the spoils of the church, and acquired Thame Abbey, built a Grammar School here. John Hampden died here in 1643. (pp. 8, 16, 54, 67, 84, 114, 179, 183.)

Watlington (1554) is an ancient market-town and has a very interesting seventeenth century market-hall in the centre of the town. A castle existed here, built by the De la Beche family in 1338, but the moat only remains. (pp. 23, 46, 80, 144, 162, 166.)

Whitchurch (946) is a picturesque village on the bank of the Thames opposite Pangbourne, with which it is connected by a wooden bridge. There are several large houses in the village, and it is not disfigured by rows of ugly cottages which have sprung up on the Berkshire side. The church has been rebuilt, but retains some interesting brass memorials. Amongst them is one to the memory of Sir Thomas Walysch, food-taster to the Lancastrian sovereigns, Henry IV, V, and VI. (pp. 27, 163, 180, 187.)

Witney (*3529) is a flourishing little market-town on the Windrush celebrated for its manufacture of blankets. A picturesque butter cross stands in the market-place. Its most interesting features are the church with its thirteenth century spire, the Grammar School founded in 1663, and the Hall of the Blanket Weavers' Company erected in 1721. The manor was owned formerly by the bishops of Winchester, who often resided here in a fortified palace erected in 1080 by Bishop Walkelin. Its name is probably derived from the Saxon word *ey* meaning an island, and *Witan*, a council, signifying the place where the Witan held its meetings. (pp. 8, 22, 30, 31, 66–8, 72, 77, 81–3, 112, 114, 163, 164.)

Wolvercote (966) is a village on the north of Oxford, near Godstow. It has a noted paper-mill which produces the paper used by the Oxford University Press. (pp. 55, 76, 85, 164, 182.)

Woodstock (*1594), a market-town on the Glyme famous for its manufacture of gloves, derives its chief historical importance from the ancient royal palace that formerly stood here, and to which reference has frequently been made. This palace has played an important part in English history. The lovers of Sir Walter Scott and the readers of *Woodstock* will remember the pranks of the loyal servant who scared the Roundhead commissioners away. Much of the story of the novel, however, is not true to history. Some parts of the old building were standing when the palace of Blenheim was in building, but the Duchess Anne ruthlessly ordered them to be pulled down. Allusion has already been made to the magnificent but ponderous pile of Blenheim. (pp. 10, 65, 67, 78, 80, 84, 88, 90, 92, 109, 127, 136, 164, 170, 173–5, 178, 182, 187.)

Wootton (1027) is a small village two miles from Blenheim Palace, where the river Glyme is joined by its tributary the Dorne. The rectory formerly belonged to Bruern Abbey and at the Dissolution of monasteries was transferred to New College, Oxford.

Wroxton, a picturesque village three miles north-west of Banbury. The Abbey, formerly a priory of Augustinian monks, is a fine Jacobean house erected in 1618 by Sir William Pope, nephew of Sir Thomas Pope, the founder of Trinity College, Oxford, who obtained the property after the dissolution of the monastic house. (pp. 122, 141, 180.)

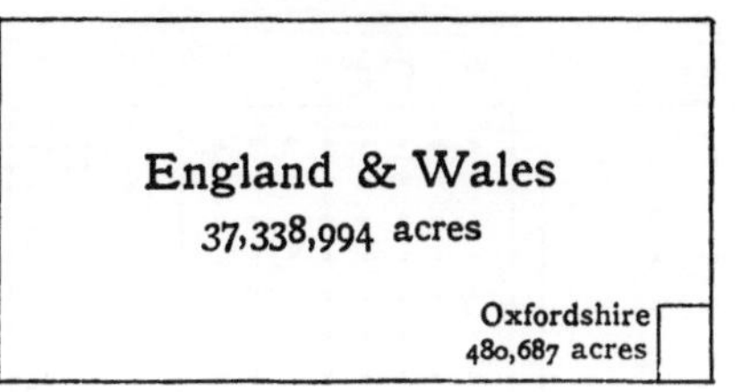

Fig. 1. The Area of Oxfordshire compared with that of England and Wales

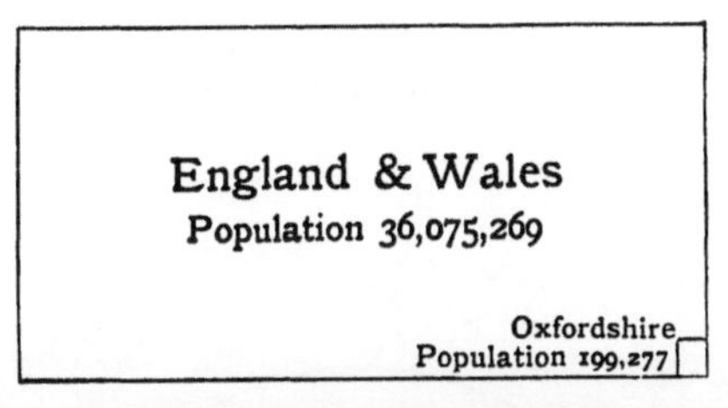

Fig. 2. The Population of Oxfordshire compared with that of England and Wales (1911)

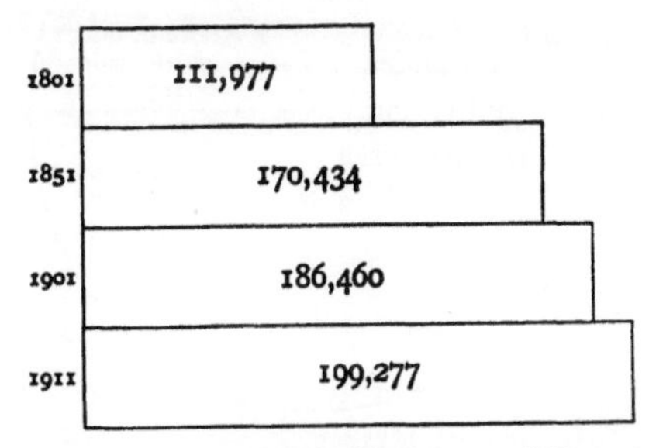

Fig. 3. Diagram showing Increase of Population in Oxfordshire

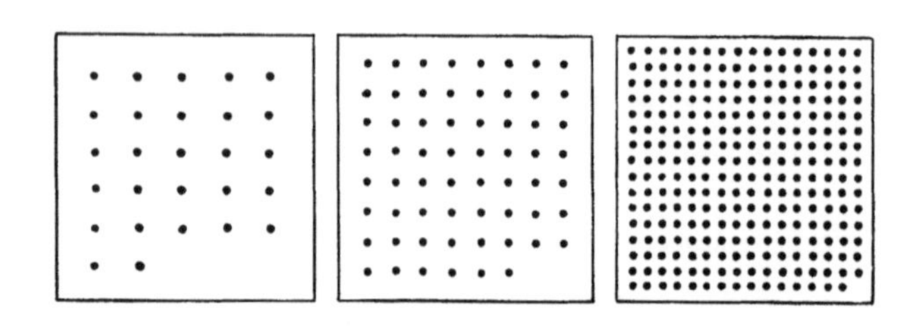

Fig. 4. Diagram showing comparative Density of Population to the Square Mile (1911)

(*Each dot represents ten persons*)

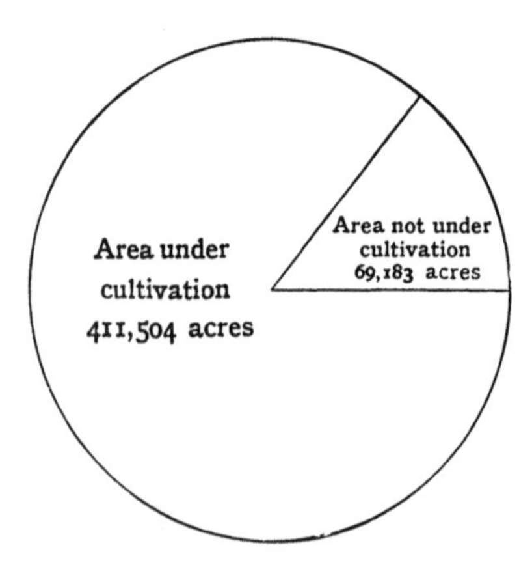

Fig. 5. Proportionate Areas of Cultivated and Uncultivated Land in Oxfordshire (1909)

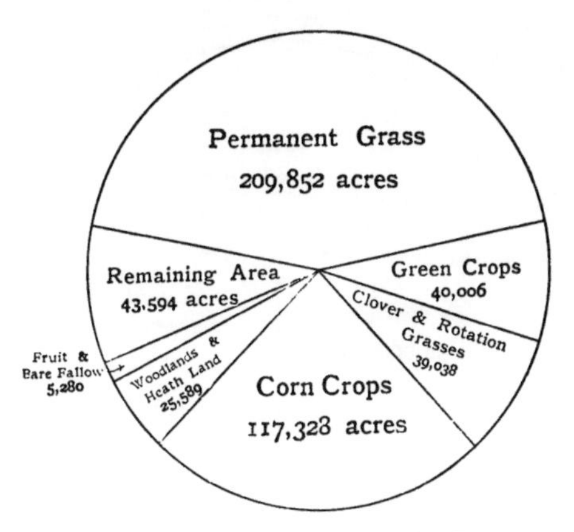

Fig. 6. Proportionate Areas of Cereals, Pasture, Crops, Woodlands, etc. in Oxfordshire (1909)

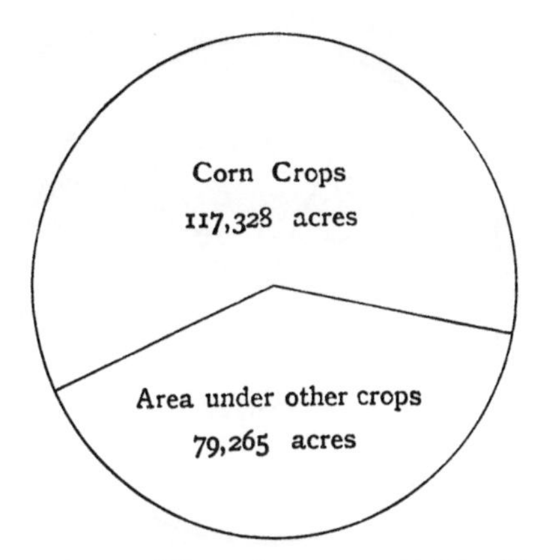

Fig. 7. Area of Corn Crops compared with that under other cultivation in Oxfordshire (1909)

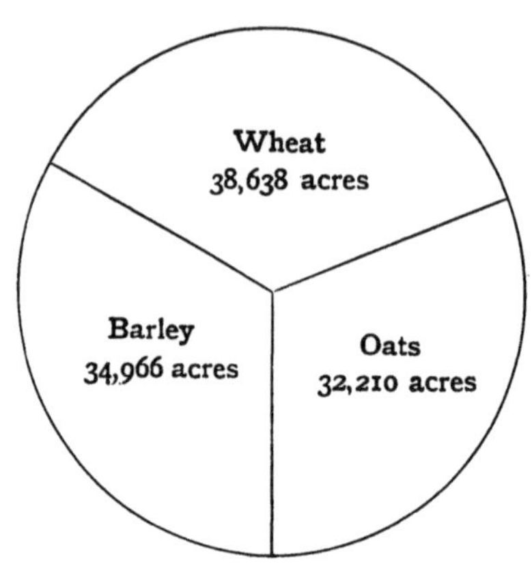

Fig. 8. Proportionate Areas of Cereals grown in Oxfordshire (1909)

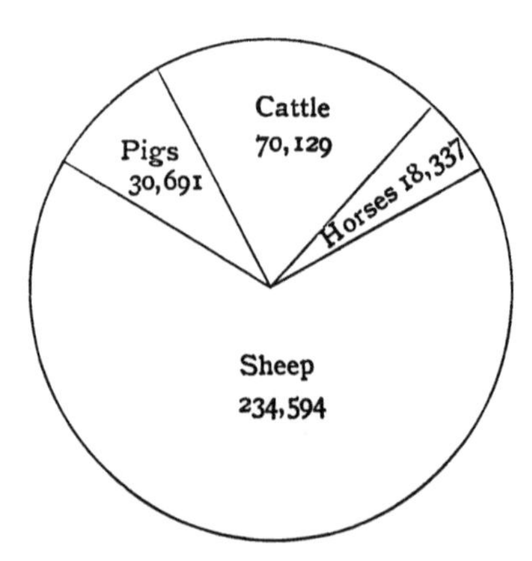

Fig. 9. Proportionate numbers of Live-stock in Oxfordshire (1909)

www.ingramcontent.com/pod-product-compliance
Ingram Content Group UK Ltd.
Pitfield, Milton Keynes, MK11 3LW, UK
UKHW042142280225
455719UK00001B/49